# European Technical Coatings Conference (ETCC 2016)



Birmingham, United Kingdom
25-26 May 2016

ISBN: 978-1-5108-2548-2

**Printed from e-media with permission by:**

Curran Associates, Inc.
57 Morehouse Lane
Red Hook, NY 12571

**Some format issues inherent in the e-media version may also appear in this print version.**



Printed by Curran Associates, Inc. (2016)

For permission requests, please contact European Technical Coatings Congress ( ETCC )
at the address below.

European Technical Coatings Congress ( ETCC )
c/o Vicentz Network GmbH & Co. KG
Plathnerstr. 4c
30175 Hannover
Germany

Phone: +49 511 9910-000
Fax: +49 511 9910-099

lena.witte@vicentz.net

**Additional copies of this publication are available from:**

Curran Associates, Inc.
57 Morehouse Lane
Red Hook, NY 12571 USA
Phone: 845-758-0400
Fax: 845-758-2633
Email: curran@proceedings.com
Web: www.proceedings.com

# TABLE OF CONTENTS

## PLENARY LECTURES



## SESSION 1: POLYMERS, PART I



## SESSION 2: CROSSLINKING & ADDITIVES, PART I



## SESSION 3: PIGMENTS & NANOPARTICLES, PART I



## SESSION 4: POLYMERS, PART I

## SESSION 9: APPLICATION PROCESS, PART I



## SESSION 10: SUSTAINABILITY & BIOBASED MATERIALS



## SESSION 11: EXTERIOR DURABILITY & CORROSION RESISTANCE, PART I

ETCC

EUROPEAN TECHNICAL COATINGS CONGRESS

# Advances in Coatings Science and Technology

FATIPEC OCCA PRA BCF BRITISH COATINGS FEDERATION

www.etcc2016.org

## How megatrends can drive development of future advanced coatings?

**Dr. Sohail Hajatdoost, Tata Steel**

Global megatrends such as urbanization, shifts in geo-political and economic powers, climate change, resource scarcity, energy and aging population are transforming different market sectors. They have impact on personal lives, society, business and economy. The megatrends that are most salient to the future of the coating industry will be discussed. Advanced coatings and surface engineering add significant value and improve the functionality of products and engineering components across the spectrum of market sectors. However, pressures from these global changes potentially create threats to existing companies or opportunities for new players. Harnessing and greater use of emerging technologies, innovation and global thinking are amongst the key success factors that ensure future prosperity of the coating industry.

Notes:

## The Roadmap to Applications of Graphene, Layered Materials and Hybrid Systems

**Prof. Andrea C. Ferrari, Cambridge Graphene Centre**

Disruptive technologies are usually characterised by universal, versatile applications, which change many aspects of our life simultaneously, penetrating every corner of our existence. In order to become disruptive, a new technology needs to offer not incremental, but dramatic, orders of magnitude improvements. Moreover, the more universal the technology, the better chances it has for broad base success. The Graphene Flagship has brought together universities, research centres and companies from most European Countries. At the end of the ramp-up phase significant progress has been made in taking graphene, related layered materials and hybrid systems from a state of raw potential to a point where they can revolutionize multiple industries. I will overview the progress done thus far and the future roadmap.

Notes:

## Session 1: Polymers, Part I

### 1.1 Silicone-based materials for coatings: Today and tomorrow

**Dr. David Pierre, Dow Corning**
*Co-Author: G.L. Witucki*

Silicon-based technologies comprise a growing portion of modern coatings. But the chemistry of Silicon-based technology, their impact on the paint's properties and advantages they can bring to any layer is still poorly known by formulators. Used solely, in simple cold-blend combinations, or copolymerized into organic hybrids, Silicon-based technologies are utilized as the primary binder in applications where heat or/and UV resistance, flexibility, adhesion are required; furthermore, a massive amount of silicon-based components are present in formulations as additives with sensitive impact on surface properties: slip, soft touch, structure, levelling and adhesion. This paper will provide a full description of the effects of silicones related to their chemical structure, and hopefully will contribute to better understand the advantages of that chemistry in the world of coatings. Through different technical examples, the paper will review and explain the impact of these versatile technologies.

Notes:

### 1.2 Ultra-low VOC waterborne alkyd coatings with solventborne alkyd coating performance

**Mr. Guy Clamen, The Dow Chemical Company**
*Co-Authors: Sue Machelski and Alan Piwowar*

Solventborne (SB) alkyd resins are widely used in industrial coatings because of their excellent gloss, adhesion, and wetting properties as well as compatibility with other resin types. In many regions, however, the consumption of SB alkyds has diminished as stringent environmental regulations have driven adoption of low VOC systems. In some instances, waterborne (WB) alkyd emulsions have replaced SB alkyd systems; however, they typically do not offer the same performance. Alternative alkyd technologies have also been introduced, i.e. high solids, water reducible alkyds, and waterborne alkyd formulations, to meet the increasingly stringent VOC regulations. Generally, though, these systems are more expensive and lower performing than their solventborne counterparts. Dow has developed technology to disperse unmodified short oil alkyd resins with minimal surfactant. Ultra-low VOC (<10 g/L), pigmented, WB alkyd coatings formulated with these short oil alkyd dispersions demonstrate comparable properties to those of conventional SB alkyd coatings including excellent gloss, adhesion, flexibility, and salt spray resistance. In addition, these WB alkyd dispersions can be blended with WB acrylic emulsions allowing optimization of dry time and early hardness development, without sacrificing gloss, adhesion, or corrosion resistance. Analysis of over 20 different acrylic emulsions shows the importance of selecting the appropriate resin to blend with these short oil WB alkyd dispersions.

The authors would like to acknowledge Martin Beebe for alkyd resin synthesis support, Robert Bills and Jamie Sullivan for alkyd dispersion support, Alan Piwowar for Analytical support, and Al Maurice, Zhenwen Fu, Jay Romick, Bob Mussell and David Carr for helpful discussions and project guidance.

Notes:

## 1.3 Small talk – of trim paints and tiny particles


### Dr. Rainer Erhardt, BASF

*Co-Author: Dr. Felix Baah*


Protection against corrosion is one of the most important functions of coatings on metal. Due to environmental considerations and tighter VOC legislation, waterborne coatings are playing an increasingly important role in the protective and industrial coatings segment. We are to confirm that acrylic (primary) dispersions are in fact very versatile to formulate anticorrosive paint. However there are different opinions in the market as to what purpose and in what environment these coatings may be used. That is why a study was done on the salt spray performance (DIN EN ISO 9227) of two styrene acrylic dispersions in zinc phosphate containing primer formulations on cold rolled steel. The dry film thickness was varied between 60 and 200 micron. In DIN EN ISO 12944 a multilayer setup of minimum 3 layers of waterborne acrylic paint with a minimum DFT of minimum 200 micron is listed for C4 medium protection level (substrate: low carbon steel). Our study confirms that also on cold rolled steel 480 h salt spray performance can be achieved using waterborne 1K acrylic paint.

Notes:

## Session 2: Crosslinking & Additives, Part I

## 2.1 The influence of crosslinking temperature and conversion on topological imperfections in crosslinked polymers


### Prof. Stuart Croll, North Dakota State University

*Co-Author: D. M. Kroll*


The formation and properties of polymer networks are usually described in terms of the chain length between crosslinks and the concentration of the network chains. However, the utility and long term durability of materials depend on flaws that are either intrinsic or develop in service. Conventional models cannot describe local or more extensive imperfections occurring during network formation and cannot quantify their effect. Molecular dynamics simulations show how networks form and that, depending on formation conditions, there may be significant numbers of loops, dangling ends and regions of low crosslinking. These effects are more pronounced in highly functional networks and those cured at high temperatures. Clearly, the heterogeneity that this implies must have an impact on the protective, barrier properties and likely longevity of those properties.

Notes:

## 2.2 In-depth study of drying solventborne alkyd coatings in presence of cobalt-free driers


### Mr. L.G.J. van der Ven, Eindhoven University of Technology

*Co-Authors: Ö. Gezici-Koç, C.A.A.M. Thomas, S.J.F. Erich, H.P. Huinink, O.C.G. Adan, J. Flapper, F. L. Duivenvoorde, L.G.J. van der Ven*


As a result of potential reclassification of cobalt-based alkyd driers, many studies have been conducted to find suitable alternatives as primary driers. A systematic investigation on the oxidative drying of solventborne alkyd coatings with commercial cobalt-free driers was performed, i.e. Nuodex Co 10 Neo, as a reference and an iron-based drier (Borchi OXY-Coat) and two manganese-based driers (DriCAT 2700F and Nuodex Drycoat). This work investigates how these driers influence the molecular network development and mechanical properties of the final films from the moment of application to full drying. High spatial resolution NMR profiling allows tracing the depth-resolved crosslinking during drying, and is used for the first time in conjunction to time-resolved ATR-FTIR and standard methods like Beck-Koller (BK) drying test, König hardness and DMA. We have found correlations between the drying pattern of driers and properties such as the drying rate, the overall and depth-resolved crosslink density and hardness development.

Notes:

## 2.3 The effect of secondary driers on drying solventborne alkyd coatings in presence primary driers


### Dr. S.J.F. Erich, Eindhoven University of Technology

*Co-Authors: Ö. Gezici-Koç, M.B. Michel, H.P. Huinink, O.C.G. Adan, J. Flapper, F. L. Duivenvoorde, L.G.J. van der Ven*


Cobalt based driers are the most used primary catalysts as they promote fast drying and lead to hard coatings. However, several studies have suggested potential reclassification of cobalt-based alkyd driers. Therefore, cobalt based driers are being replaced by acceptable alternatives. Primary driers are used in combination with secondary driers to enhance homogeneous drying of the film. During the evaluation of new driers, coating formulators aim a high drying speed and hardness development. However, each combination of primary and secondary driers leads to a different oxidative drying pattern resulting in variations in the crosslink density and hardness development. We investigated the curing behaviour of secondary driers, Ca and Zr, using high spatial resolution NMR set-up. We observed increased drying speeds and higher crosslink densities when adding secondary driers. Totally unexpected, we observed metal transition of manganese during the drying process.

Notes:

## Session 3: Pigments & Nanoparticles, Part I

### 3.1 Smart fillers for smart coatings

**Prof. Maria Zielecka, Industrial Chemistry Research Institute**

*Co-Authors: Elzbieta Bujnowska, Regina Jeziórska*

It is well known that smart coatings make smart products. One of the effective ways to obtain smart coatings is incorporation of multifunctional, smart fillers to polymer matrix. The effect of multifunctional fillers on final properties of polymer composites will be presented based on the following types of fillers:

- Non-functionalised and functionalised silica nano-spheres
- Inorganic fillers applied for silicone ceramisable rubber

Using synergy of very rich silicon compounds chemistry and the possibility of precise steering of sol–gel process original methods of manufacturing silica nano-spheres with unique properties were developed. Silica nano-spheres containing immobilized silver or copper nanoparticles of narrow particle size distribution were also synthesised. All these types of silica nano-spheres can be applied as nano-fillers for polymer nano-composites in the purpose to achieve better thermal, mechanical, barrier and antimicrobial properties.

The other example of multifunctional fillers application is polymer composites ceramizing at high temperature arising during fire.

Notes:

### 3.2 Snail trails and colour brilliance – how a new composite technology tackles diverse challenges

**Mr. Lenine de Sousa Gomes, Wacker Chemie**

The aim of this work was the development of a dispersion based on composite technology that enabled the formulation of facade paints with higher performance levels than standard technologies involving acrylics and vinyl acetate copolymers. Particular points of focus were dirt pick-up, color retention and snail trails. The recently introduced PRIMIS® Composite Technology based on pure acrylics and inorganic silica can be used to reduce snail trails in facade paints, as well as being an optimal host for incorporating critical organic pigments and reducing dirt pick-up. In comparison to other composite technologies, a more homogeneous distribution of silica in the dried paint film can be achieved with PRIMIS® Composite Technology through the combination of silane chemistry and silica sol. Better silica distribution improved the usual problems in facade paints based on composite technology, for example formation of cracks and high water uptake.

Panels coated with PRIMIS® Composite Technology based paints were exposed in an outdoor weathering station located in Dubai (UAE), where sand storms are very challenging for colored facade paints. The exposed panels had much less dirt pick-up than panels coated with paints based on standard technologies. Numerous simulated and real outdoor weathering tests conducted in Burghausen (Germany) showed that PRIMIS® Composite Technology is an excellent alternative for colored facade paints as regards color development and fading, whether with inorganic or organic pigments. Snail trails can be related to the migration of salts (leaching) through gentle rain or dew from the painted substrate and from the paint itself to its surface, disturbing the visual appearance of the facade paint. Surface evaluations performed in our laboratories showed that paints based on PRIMIS® Composite Technology displayed better leaching resistance than conventional paints elaborated with acrylics and other technologies.

Notes:

## 3.3 Developing a complete cool colours solution for the coatings industry

**Mr. Daniel Lladó, Nubiola (A Ferro Company)**

*Co-Authors: Patrícia Pérez, Juan Manuel Somoza, Jordi Fernández*

It is well known that black or dark surfaces heat up under the sun, while white surfaces remain cool. This heat build-up is *related to* the level of absorption of Near Infrared (NIR) wavelengths from the sun and one way to measure it is by Total Solar Reflectance (TSR), expressed as a % of the light reflected by the studied surface.
Those pigments allowing high reflection of Near IR from the substrate and thus a high TSR are referred to as cool pigments and are usually associated to Complex Inorganic Color Pigments. Ferro Corporation is supplying the market with this kind of high performance pigments since long and their cool properties are well known in the market.
However, other pigment chemistries not so widely used for cool coatings are also suitable for this application, such as high performance Ultramarines and Zinc Ferrites manufactured by Nubiola, recently acquired by Ferro Corporation. One of the immediate synergies to offer to the Coatings industry by the united Ferro-Nubiola forces is the development of a complete color solution for cool applications in a wider spectrum of systems with different performance requirements.
In this paper the cool performance of Ultramarines and Zinc Ferrites is proved and the color and performance synergies with Complex Inorganic Color Pigments in the Blue, Beige and Black color space are explored.

Notes:

# Session 4: Polymers, Part I

## 4.1 Advanced isosorbide-based wood coatings - polyurethane acrylic hybrids

**Dr. Claudia Schirp, Fraunhofer WKI**

*Co-Authors: Steven Eschig, Dagmar Liphardt*

Biobased renewables are of high interest as they may substitute petro-based chemicals in the long-term future. Many bio-based, alternative monomers are already taken into account, both in scientific researches and in commercial use. Most of them are excellent substitutes regarding flexible and elastic properties of wood coatings. In contrast, isosorbide as a bicyclic diol, is considered to be a suitable monomer, bringing hardness and impact resilience to the resulting polyester-based polyurethane dispersion. These dispersions can be used as binders for wood coatings. In addition to the results shown at ETCC 2014, acrylic polyurethane hybrid resins will be presented. The research was focused on the demonstration of chemical bonding between the acrylic and the polyurethane part. Therefore, hybrids and blends were synthesized and compared by different analytical methods. Finally, the polymer composition was optimized and the resins were formulated to test coating film properties for various applications.

The results were partly gained within the EU project „NEXT1KOAT". The authors would like to thank for support in lab work to Sandra Hofmeister and Simon Niegebar, the project partners INSPIRALIA (Spain), CETEM (Spain), WIC (Slovenia), BPF (UK), MFMO (Malta), HELIOS (Slovenia), ISOLACK (Spain), RIOMADER (Spain) and BOD AYRE (UK) for kind assistance and the European Community's Seventh Framework Program managed by REA-Research Executive Agency (FP7/2007-2013) for funding under grant agreement n° 304998.

Notes:

## 4.2 Modified starch as potential water-based binder system for interior coatings

**Ms. Christina Gabriel, Fraunhofer IPA**
*Co-Authors: H. Wetzel, S. Radosta, W. Vorwerg*

With its worldwide availability and its low market-price native starch provides the potential for the use as a bio-based material. However, in its native form it is highly hydrophilic, not cold-water dispersible and only forms brittle films. So, our work aimed at the chemical modification of degraded starch for the application in the coating sector as binder, with the focus on paints and varnishes. For this purpose, cold-water dispersible hydrophobized starch esters and ethers with high solid contents (≥ 30% (w/w)) and viscosities between 100 and 1000 mPa·s should be synthesized. Additionally, they should form continuous films and show adhesion to different substrates.

As starting product a degraded potato starch with a molar mass of $1.18 \cdot 10^5$ g/mol was used and modified by carboxylic acid anhydrides ($C_2$-$C_4$) and carboxylic acid vinylesters ($C_6$) in the case of the esters. Epoxy alkane derivatives were used for the synthesis of hydroxyalkyl starch ethers ($C_3$-$C_4$, $C_6$). Beside the type of substituent and the chain length, also the degrees of substitution (DS) and molar substitutions (MS) were varied from 0.05 to 1.1.

The modification led to cold water-soluble or dispersible products with applicable concentrations between 30 and 45% (w/w) and viscosities in the range of $10^2$-$10^3$ mPa·s (100 $s^{-1}$). The most samples had a newtonian flow behaviour and showed viscoelastic properties with higher values for the loss modulus (G"). Continuous films were observed with starch esters in the DS range of 0.11-0.81 and in the case of the starch ethers with MS values of 0.77-0.91. Wet and dry layer thicknesses of 25-50 µm and 7-12 µm were obtained (doctor blade: 100 µm). In the cross cut test the esters had clearly better results with values of 0 and 1, in contrast to the tested ether with values of 4.

The compatibility tests with 3 commercial binders showed that two Acronal binders (A 754 & S 559) (42%) showed compatibility with the modified starches (58%).

According our experiments and especially based on the results of the film formation and the cross cut test the starch esters and not the ethers could be the preferred class for an application in the coating sector. Additionally, the results showed that the starch esters could be an alternative to petroleum-based polymers as binder in paints and varnishes potentially.

Notes:

## 4.3 Unveiling alcohol resistance in waterborne coatings

**Dr. Silfredo Javier Bohorquez, Nuplex**
*Co-Authors: Andrew Teasdale, Jaap Akkerman and Dirk Mestach*

Solventborne polyurethanes binders have been the main choice for wood coatings for many years due to their outstanding performance for chemical resistance. Nevertheless, the worldwide awareness about the sustainability and the different factors contributing to indoor air quality is the major cause for the shift to waterborne technologies. Waterborne coatings for wood applications need to fulfil several properties in order to perform at a similar level to traditional coatings. Among these properties: Fast drying, sandability, stackability and chemical resistance. Nowadays, alcohol resistance is one of the most demanded properties of waterborne 1K systems for furniture applications. In that sense, this work reveals the re-engineering of a waterborne binder based on a deep evaluation of alcohol attack based on microscopy analysis. The relationship between the macro and micro (AFM, SEM and TEM) phenomena led to the design of glossy and matte 1K systems with excellent alcohol resistance.

Notes:

### 4.4 Why blends of acrylic and polyurethane dispersions do not always form clear films

**Dr. Wolfgang Gaschler, BASF**
*Co-Authors: M. Conrad, Prof. B. Strehmel, S. Kirsch, S. Enck*

Blending polymer dispersions of different kinds to tailor coating properties is routine in the coatings industry. This paper explains why mixtures of polyacrylate and polyurethane dispersions do not always form a transparent film. The root cause of turbidity is a mismatch of the respective refractive index of the polymers. Besides refractive index differences, the film morphology and domain sizes influence the degree of turbidity. We can demonstrate with the help of confocal laser scanning microscopy that inhomogeneity in the polymer films has its origin in the wet state due to an uneven distribution of the latex particles. Polyacrylate-polyurethane hybrids circumvent the problem of film turbidity by very small domain sizes and come with the advantage of superior mechanical properties.

Notes:

### 5.1 New water dispersible HDI-polyisocyanates for 2K-PU WB Coatings

**Mr. Junya Tsujita, Asahi Kasei Corporation**
*Co-Author: K. Nakajima*

Due to VOC regulations the number of waterborne coating formulations has considerably increased. To meet the various requested properties of different application fields it is necessary to adjust i.e. hydrophilicity, physical drying performance, etc. of a hydrophilic polyisocyanate. Hence, the Asahi Kasei Chemicals Corporation is developing continuously new hydrophilic HDI-polyisocyanates for 2K-PU waterborne coatings.

This presentation introduces two new and innovative HDI-Polyisocyanates. One hydrophilic Polyisocyanate can reduce the physical drying time of 2K-PU waterborne coatings compared to using standard hydrophilic Polyisocyanate significantly. Therefore this paper discusses the concept of multi-functional technology in a first part.

The second part of this presentation focusses on the controlled hydrophilicity technology and introduces a hydrophilic Polyisocyanate which provides easy incorporation behaviour. Especially formulations aiming for hand-mixing only, benefit at great length by using this new invention.

Notes:

## 5.2 Novel acrylic epoxy hybrid coatings for metal protection applications

### Mr. Guy Clamen, The Dow Chemical Company

*Co-Authors: Andrew Trapani, Sandrine Sinsoulieu, Zhenwen Fu*

Two components waterborne epoxy coatings have been used for some time for protective coatings. However, the current waterborne epoxy coating systems are generally slow curing and the standard epoxy-based systems are not inherently UV light stable. A new waterborne Acrylic Epoxy Hybrid (AEH) dispersion technology has been developed to address these issues. The AEH can be formulated as a stable 2K coating by mixing pigments and fillers with waterborne polyamine hardener or carboxylic based acrylic hardener as part A and the AEH as part B. The 2K formulation stability is controlled by the hybrid and hardener surface charge reflected in zeta potential values. With an appropriately hydrophobic hardener, the 2K coating can achieve much better corrosion resistance than 1K acrylic and 2K acrylic/liquid epoxy resin systems and similar to conventional 2K waterborne epoxy systems. This 2K coating dries faster than a traditional 2K WB epoxy but has much longer pot life and excellent dilution stability.

Notes:

## 5.3 A siloxane for amine crosslinked systems

### Dr. David Pierre, Dow Corning

*Co-Author: G.L. Witucki*

2K Epoxy-Amine systems are well-known in the Protective coating market. A new generation of amino functional Silicon-based materials is proposed to formulators, combining the reactivity of the amine and the benefits of siloxane. This paper will review the different effects of an amino siloxane resin used as main or co-binder in such a system. The siloxane backboned binder will be described as an impactful component on the final properties of the paint: heat and chemical resistance, flexibility, adhesion... The effects of this material will be described through different examples and descriptions.

Notes:

## 5.4 Strategies for thickening vinyl binder paints with HEURs for optimal application properties

**Mr. Pol Storme, The Dow Chemical Company**
*Co-Authors: Gerry Valdez, Ant Van Dyk, Mark Langille*

Vinyl binders, such as vinyl acrylate (PVA) and vinyl acetate ethylene (VAE), account for a significant portion of all architectural coatings binders sold globally. These paints are commonly thickened with non-associative cellulosics, such as hydroxyethyl cellulose (HEC). HEC is a low cost thickening option but can yield paints with poor flow and leveling. Good flow and leveling are critical for brush marks and roller patterns to dissipate after application of the paint to the substrate. However, sag resistance is another important consideration. Drips or curtaining of the paint negatively impact the final film appearance. Flow and leveling require that the paint have a low viscosity after application whereas sag resistance requires that the paint have a high viscosity after application, seemingly opposite rheological needs.

Associative thickeners, such as hydrophobically-modified ethylene oxide-based urethane (HEUR) rheology modifiers can provide an excellent balance between sag resistance and flow/leveling needs. However, HEURs are typically not used in combination with vinyl binders because they can have low thickening efficiencies due to the hydrophilic nature of the vinyl binder surface. We demonstrate here that some of the desirable application properties of HEURs can be brought into vinyl binder paints by replacing a small portion of the HEC with HEUR. When compared to an all HEC thickened paint, we find that replacing 10-30% of the HEC with a high efficiency HEUR can simultaneously improve the sag resistance and flow/leveling of the paint. This can be accomplished without a significant increase to the total amount of rheology modifier used in the paint. We show how this strategy can be used to enhance the sag/level balance in VAE and PVA paints, which in turn can result in improved application properties by reducing roller patterns and achieving a smoother surfacefinish.

Notes:



## 6.1 Graphene: A multi-functional nanomaterial for the coatings Industry

**Ms. Lynn Chikosha, Applied Graphene Materials**
*Co-Author: Tim von Werne*

Graphene in its purest form is usually thought of as being a single sheet of carbon atoms, although the term graphene is generally accepted to apply to materials that are up to ten layers thick, the properties of which can be different to those of pure single-layer graphene. Graphene possesses an unsurpassed combination of electrical, mechanical and thermal properties, which gives it the potential to replace existing materials in a wide range of applications and, in the long term, to enable new applications.

Graphene has been incorporated into two model coating systems, a two pack polyurethane clear coat and a two pack epoxy based coating. The coatings were then tested for a wide range of properties including colour, gloss, corrosion resistance, scratch resistance and tensile modulus. The initial results demonstrate that low loading levels of graphene nano-platelets substantially enhance the scratch resistance and ultimate tensile strength of the coatings, and critically, with minimal impact on transparency or colour. The preliminary results from our test work are very interesting and show that well dispersed graphene nano-platelets offer the potential for significant property improvements in polyurethane coatings. This opens up the route to further development for commercial applications in numerous high performance paints and coatings.

Notes:

## 6.2 CNTs and antifouling coatings: Sustainable synthesis, safe-by design and release assessment

**Ms. Vasileios Markakis, National Technical University of Athens**

*Co-Authors: Elias P. Koumoulos, Costas A. Charitidis, Aikaterini-Flora Trompeta*

Addition of carbon nanotubes (CNTs) in antifouling coatings exhibits promising antifouling results due to the hydrophobic nanostructured surface obtained. However, in order this novel technology to take off, is important to study all the environmental burdens that occur, not only due to the substances the antifouling coatings contain, but also due to the synthetical routes employed for the production of CNTs. Life cycle analysis (LCA) paves the way for a sustainable synthesis of novel coatings that exploit nanomaterials for improved and extraordinary properties. In this study, a step-by-step LCA is presented that includes the assessment of synthesis, functionalization and incorporation of CNTs into the coatings, as well as the impacts of the CNT–composite coatings. In this sense, three major steps of LCA have been implemented. At first, the life cycle of CNTs synthesis has been studied, followed by a life cycle analysis on the necessary surface functionalization process, needed for the improvement of compatibility between CNTs and the coatings matrix. Some first results from the LCA study showed that in the case of CNTs synthesis, energy consumption is the major environmental contributor, while for the functionalized CNTs, the chemical reagents and their emissions seem to be responsible for the environmental impacts. The above, have been incorporated in the LCA of an indicative paint, blended with the functionalized CNTs. The results showed, that in the case of copper(I) oxide substitution by CNTs in the antifouling paint, the total environmental load seems to be decreased. This holistic study has been realized in the framework of the safe-by-design concept, in order to produce commercial paints that will not be only end-user friendly, but also will offer occupational safety. Taking into account this concept, the whole procedure has been assessed in terms of environmental impact, filling a gap in literature about the consequences of nanomaterials in commercial products.

Acknowledgements

This research is supported by the FP7-OCEAN-2013 Project "Low-toxic cost-efficient environment-friendly antifouling materials" (BYEFOULING) under Grant Agreement no. 612717

Notes:

## 6.3 Thermal behavior of epoxy-polyester powder coatings modified with nanofiller

**Ms. Joanna Trzaskowska, Industrial Chemistry Research Institute**

*Co-Authors: J. Kozakiewicz, I. Ofat-Kawalec, I. Legocka, H. Kuczynska*

In this study thermal behavior of epoxy-polyester powder coatings both unmodified and modified with silicone-containing nanofiller obtained through spray drying of aqueous silicone-acrylic core-shell dispersions were investigated using Differential Scanning Calorimetry (DSC) and Thermogravimetric Analysis (TGA). The effect of different temperature regimes on the coating characteristic features (glass transition temperature, melting temperature, onset cure, curing temperature, heat of fusion) was also studied using DSC method. The experiments revealed that nanofiller addition to epoxy-polyester powder coatings affected polymer phase transition temperatures and thermal stability.

Notes:

### 6.4 The effects of Nano-SiO2 particles on the rheological and anticorrosion properties of an automotive electrocoat

**Prof. Zahra Ranjbar, Institute for color science and technology**
*Co-Authors: M.Bakhtiary, S.Moradian*

It has been attempted to modify the rheological and anticorrosion properties of an automotive electrocoat using nano-silica particles in order to improve the edge coverage at complex geometries. Different weight fractions of nanoparticles were added to the electro-coatings. The rheological properties of the electro-coatings were studied by RMS and Brookfield viscometer. The anticorrosion performances of the coatings were studied via EIS and salt spray tests. Results showed that addition of certain values of nanoparticles to the electro-coating could improve its rheological and anticorrosion properties significantly. The edge coverage of the coating significantly enhanced in presence of the optimum nano-silica loadings. The curing behaviour of the electro-coating was influenced in presence of nano-silica particles. The cross-linking density and Tg of the coating were reduced in presence of high particles loadings. The roughness and gloss of the electrocoat was influenced in presence of nanoparticles.

Notes:

## Session 7: Characterization & Processing

### 7.1 Understanding open time: Fundamental mechanisms from particle motion to paint workability

**Dr. Antony Van Dyk, The Dow Chemical Company**
*Co-Author: Lin Wang*

Latex paints have many advantages over solvent-based paints, such as drying by evaporation of water and rapid development of resistance properties. Practical application entails spreading, removal, redispersion, redeposition, and respreading of paint on an area over time. Conditions and substrate govern the evaporation rate; even with high VOC paints, environmental factors of temperature, relative humidity, airflow, and substrate porosity, dominate the rate of drying. Capillary transport becomes more important at later stages of drying, depending on factors such as PVC, particle sizes and distributions, interfacial tensions and contact angles, rheology and hydrogen bonding additives. Microstructure develops with the formation of packed surface layers that grow in vertical extent and compaction, much like the formation and aggregation of ice floes on water near a phase transition. Smooth, defect free, application can be challenging, particularly under dry, warm, and windy conditions, where thin patches and edges tend to dry before achieving uniform coverage. Open time characterizes the ability to maintain a wet edge and spread the paint uniformly.

Various test methods seek to quantify open time (ASTM-D7488, BK recorder ASTM D5895, ISO-9117, brush or roller application, six panel door, plus variations and informal methods). Skill and technique interact with rheology and open time evolution. Open time is typically thought of as some defined time after which paint fails to be workable. The reality is more complex as application properties evolve and deteriorate during painting of an area. We explore the mechanical and rheological properties of bulk coatings and surface layers to give a better understanding of open time fundamentals. Diffusing Wave Spectroscopy, tribology, rheology, Molecular Dynamics simulations of latex and additives, gloss and refractive index, image analysis, and modeling are to used characterize this evolution to provide insight into better characterization and control of open time.

Notes:

## 7.2 Why test inks cannot tell the full truth about surface free energy

### Mr. Nicholas Hearn, Krüss

*Co-Authors: Ming-Jin, Thomas Willers*

A solid's surface free energy plays a determining roll in the performance of both coating and bonding processes that are used today in a wide range of industries. Determining a solid's surface free energy is useful not only in checking that the surface properties are compatible from the point of view of wettability, with the coating or adhesive but also to check that the functional surface properties of the finished coating have been achieved. One example of this would be that of super hydrophobic surfaces where self-cleaning properties are achieved through the lotus effect.

For many materials extensive pre-treatment such as cleaning, surface roughening or plasma activation are applied prior to coating or bonding. All of these treatments have one major goal in common: improving the wettability of the coatings and adhesives on surfaces. To monitor the efficiency of these processes, the surface free energy (SFE) of the substrate is measured. In many cases dyne inks are used to determine the total SFE following the assumption that a surface having a SFE value above a certain threshold is sufficiently treated for the subsequent coating or bonding.

However, the SFE is more than one single value and its separation into polar and disperse components is essential if wetting and long term adhesion are to be characterised. In contrast to dyne inks, contact-angle measurements determine the polar and disperse contributions to the SFE.

In a thorough experimental study, we determined the SFE of various materials using different types of dyne inks (yellow ethanol and blue formamide based ones) and contact-angle measurements. The investigated materials range from polymers (polyamide, polypropylene, polyethylene, polytetrafluoroethylene, polyvinyl chloride, and polyethylene terephthalate), glass, silicon wafer, mica, and graphene, to metals like aluminium. In addition, we tested some materials before and after plasma treatment. Results obtained with test inks and contact-angle measurements will be compared illustrating the advantages and drawbacks of either technique. We explicitly explain why for some materials test inks and contact-angle measurements yield different results. For that purpose we determined the polar and disperse components of the test inks. Finally, we present a case study on coating in the automobile industry, where we illustrate how contact angle measurements involving the substrate and the liquid coating can provide important information about the work of adhesion and interfacial tension between them. As these influence short-term bonding and long-term adhesion respectively, they are essential in fine-tuning the coating process.

Notes:

## 7.3 Future dispersion technology in lacquer and paint production

### Dr. Hans-Joachim Jacob, Ystral

The presentation gives a summary and explains details of the newest developments and latest experiences with Inline Dispersing Machines, which have started to displace the High Speed Dissolvers in the lacquer, paint and printing ink production. In many cases the agitated media mill is substituted as well.

Technical parameters determining the dispersion effect and quality like shear gradient, specific dispersion power, energy density and specific liquid surface during powder addition are up to 10.000 times higher compared to a HSD. This leads to a much more intensive dispersion and better qualities in a shorter time.

Typical economic effects are:

- Reduction of production costs to below 10 %
- Reduction of production times to below 20%
- Reduction of energy costs to below 30%
- Production with just 40% of the operators up to 400% more product
- Minimum space requirement
- Lower investment costs
- Cost savings in raw materials

Notes:

## Session 8: Exterior Durability

### 8.1 Test precision of the EN 927-6 artificial weathering test for wood coatings

**Mr. Martin Arnold, Empa**

An estimate of test precision (repeatability and reproducibility) is important for a correct interpretation of weathering performance data. However, many test standards still lack specific precision data, because they require extensive and costly interlaboratory studies. In the frame of the currently running EU-project SERVOWOOD, test precision has been assessed for the fluorescent UV artificial weathering according to EN 927-6. In co-operation with CEN/TC 139/WG 2 "Coating systems for exterior wood", this was accomplished by a comprehensive round-robin exercise between the involved research labs, including both common and home sourced wood panels coated with a range of model coatings and exposed in sequential runs by each participant.

In agreement with the precision data already established for natural exposure according to EN 927-3, the performance data of the semi-transparent Internal Comparison Product (ICP) are regarded typical for this test method and used as basic estimates to calculate repeatability and reproducibility limits. However, test precision may vary with coating systems and assessed material properties.

Calculations were based on 3 sequential exposures of 2016 hours duration in 5 different laboratories with sets of 3 replicate wood panels coated with 3 coats of the ICP. Panels were produced by the participants individually, applying a common batch of ICP to pine wood sourced locally. Performance assessments were done by different operators and using different measuring devices at the 5 laboratories.

Based on these results a proposal for a precision statement was submitted to CEN/TC 139/WG 2 to be included in a future revision of EN 927-6.

Acknowledgement:

The SERVOWOOD project (www.servowood.eu) receives funding from the European Union Seventh Framework Programme (FP7/2007-2013) under grant agreement FP7-SME-2013-606576. The overall aim is to assess and improve the capability of short term exposure tests to predict the performance of exterior wood coatings under field conditions and to develop European standards that will facilitate the prediction of service life. Contributions to the project from all consortium members are acknowledged.

Notes:

### 8.2 Can accelerated weathering predict influence of TiO2 on outdoor durability of white topcoats?

**Ms. Annemie Verhaege, Chemours**

*Co-Authors: Dr. Michael P. Diebold*

There is much debate about the accuracy of accelerated durability tests in predicting the real outdoor durability of paint systems. However, these accelerated tests are often used in the paint industry as a screening method for ingredient selection. International standards like Qualicoat or GSB also demand these accelerated tests.

In this paper the impact of different $TiO_2$ types on the durability performance in white powder and coil coat formulations will be discussed. Our main interest is in determining the ability of these accelerated tests to adequately predict exterior performance. The study demonstrates the risk of using accelerated data in view of predicting the final durability class.

There are many paint components that can affect the durability of a paint film. The influence of individual paint components on gloss retention and color fade can be complex. For our study we chose to focus on the effect of $TiO_2$ grade, not only on overall film durability but also on the relationship between accelerated and natural outdoor exposure results. Paints were exposed in QUV-B, Xenon Weather-o-meter and Florida, and the results were reviewed according to Qualicoat and GSB classifications.

We found that for the PE powder system, neither QUV-B nor Xenon Weather-o-meter exposures were able to differentiate between $TiO_2$ grades, while outdoor Florida testing did reveal that some paints were significantly less durable than others.

Two different commercial PE coil resins were investigated in a similar way. The predictive power of the accelerated tests versus Florida was again questionable. The overall durability performance of the PE resins was also not completely in line with expectations. In addition, in some cases we saw contradictory results, where some grades with low gloss retention had good color retention.

Acknowledgements to DSM for preparing the powder coat panels and to the Chemours Technical Marketing group for preparing the coil coat panels.

Notes:

## 8.3 Prediction of colour development of coated wood surfaces during weathering and maintenance

### Mr. Boris Forsthuber, Holzforschung

*Co-Authors: Monika Ecker, Gerhard Grüll*

The aim of the present study was the modelling of the photo-induced discolouration of coated wood during simulated sunlight exposure. Additionally, the influence of applied maintenance coatings on the resulting colour of the coated wood surfaces was quantified. For that purpose, wood samples of four wood species were coated with three waterborne semi-transparent coating systems with varying pigmentation. After 24 h and 48 h of exposure, maintenance coatings were applied on parts of the exposed coated wood samples.

Best results concerning the modelling of discolouration were obtained, using the CIE XYZ colour space with an exponential decay fitting function. The resulting model was further improved by taking the differential colouration of early- and latewood into account. The results regarding the discolouration by application of semi-transparent maintenance coatings showed a shift of the colour model, which is determined by the degree of pigmentation. The resulting models could be of great use to aid customers in predicting and visualizing the resulting colour tone of maintained coated wood building elements.

Acknowledgments:
The project SERVOWOOD (www.servowood.eu) receives funding from the European Union Seventh Framework Programme (FP7/2007-2013) under grant agreement FP7-SME-2013-606576. Contributions to the project from all consortium members are gratefully acknowledged.

Notes:

# Session 9: Application Process, Part I

## 9.1 Led-cured self-replenishing hydrophobic coatings based on interpenetrating polymer networks

### Mr. Yujing Zhang, TU Eindhoven

*Co-Authors: F. Karasu, C. Rocco, C. Croutxé-Barghorn, X. Allonas, L. G. J. van der Ven, R. A. T. M. van Benthem, A. C. C. Esteves, G. de With*

The concept of self-restoring hydrophobicity on polymeric coatings, by replenishing the functional groups on the damaged surfaces, is of high interest for industrial applications. The proof-of-principle of self-replenishing hydrophobic coatings was reported for Polyurethane-based networks with fluorinated dangling chains, which can re-orient towards the surfaces created upon damage. However, the high temperature and long-time curing procedure, low hardness are identified as the main hurdles for applications.

In this work, we investigate the self-replenishing behaviour of LED-cured IPN-based hydrophobic coatings. First, we demonstrate the feasibility to prepare photo-cured self-replenishing hydrophobic coatings with PEG-diacrylate-based networks via UV-curing. Based on this acrylate system, an acrylate/epoxide combination has been chosen to form IPN-based coatings cured by LED (daylight) with two different Tgs. These systems allow combining the flexibility of the acrylate, which is necessary to offer enough mobility for dangling chains reorientation; and the stronger mechanical properties of the epoxy, to improve the hardness.

Notes:

## 9.2 Cheap and robust 3D printed applicators for use in robotic high-throughput film application


### Mr. Peter G. C. A. van den Berg, Nuplex

*Co-Authors: Jaap M. Akkerman, Eldert J. Valk and Aschwin van der Horst*


Film application of paint is normally applied by a draw-down bar, also known as "bird" or "doctor blade" applicator, with a well-defined gap size. Automation of this process by using such a metal applicator yields a difficult and costly cleaning process especially when applying waterborne coating. Since the development in 3D printing technology has rapidly evolved the technique is now a days a bog-standard. Because of the latter, design, manufacturing and use of low cost 3D printed applicators with its 3D printed supports used in robotic high-throughput film application becomes within reach. In our laboratory we have developed such applicators with well-defined and reproducible gap sizes for the use in our automated application robot. All ins and outs of the automated film application development by 3D printing technologies will be evaluated in this presentation.

Notes:

## 9.3 New continuous production method of large area EL-Systems


### Dr. Carsten Cornelissen, InovisCoat


InovisCoat GmbH has developed a completely new technology for large area electroluminescence systems for different types of substrates like films, textiles and papers. Electroluminescent devices are usually produced by screen printing of pastes made from organic solvents. InovisCoat produce EL-systems by applying continuously, simultaneously and mixing free water-based multilayer-systems.

Inoviscoat is using a cascade coater to transfer a coating package – e.g. an electroluminescence multilayer system - consisting of up to nine differently functionalized layers without intermixing of the adjacent layers. The process and the resulting products have the following advantages: 1. production of large formats possible (roll-to-roll), 2. environmentally sustainable, 3. economical production (multiple layers in one step), 4. high precision and productivity, 5. highly flexible films.

The electroluminescent systems are manufactured in various colours and can be tailored and combined with any product. It will be integrated into security signage, protective clothing, ambient light systems, advertising and other markets.

Notes:

## 10.1 Bringing the world to life: the BCF sustainable development policy, 7 years on

**Mr. Tony Mash, British Coatings Federation**

In 2009, the British Coatings Federation (BCF) published a landmark document that defined Sustainable Development for the UK-based paint, wallcoverings and printing ink industries, and launched a bold set of commitments to which its members were invited to subscribe. This document catalysed similar initiatives across Europe and beyond.

Seven years later, the BCF has collated a wide number of case studies which demonstrate that its policy document has been supported with energy, enthusiasm and creativity. The presentation features many impressive examples of Sustainable Development which have been incorporated into company performance, not only addressing environmental aspects of Sustainability but also economic and social impact pillars as well.

However, the BCF recognises that Sustainable Development is a continuous process and there is always room for further improvement. It is argued that Life Cycle Thinking and the Circular Economy will play an increasingly important part in the thinking of the coatings industry in the years ahead.

History has shown that co-operation along the supply chain is an essential ingredient for sustainable development to break new ground and develop improved methods of working. The coatings industry interfaces with both material supply industries and the applicator/consumer, and is therefore well placed to lead the whole supply chain towards a more sustainable future.

Tony Mash, who was CEO of the BCF at the time of this document's launch, will summarise what has been achieved to date by all aspects of the UK coatings industry and propose a set of key priorities for further development and improvement over the next seven years.

Notes:

## 10.2 The importance of energy pricing in building a sustainable future for the coatings industry

**Mr. Jonathan Bourne, Nexant**
*Co-Author: Tzyy Lok*

The drivers of technical change in the coatings value chain are well established and the industry continues to evolve in response to the challenges of lower growth in maturing markets, energy cost, materials efficiency and legislation; sustainability is the new buzz of this century but incorrect assumptions about future energy costs could lead to higher risks and inefficient allocation of precious technical resources when planning R&D projects.

Sustainability addresses the way in which we innovate and grow, taking due account of economic, social and environmental impacts. As a consequence of the growing use of hydrocarbon fuels, concerns over global warming and occasionally steep rises in prices, we have seen a surge of interest in 'green' chemistry over recent years i.e. the search for new or replacement materials sourced from bio-based origins and processes. The use of bio-based raw materials in paints is not new, as natural ingredients have been used since they were first invented; however the developments in synthetic chemistry over recent decades have enabled many new materials, with superior performance properties, to be produced. It is little wonder then that modern research efforts have sought to replicate or improve upon these materials using 'white biotechnology' as a new means of synthesis.

For invention to turn into innovation the economic outlook must be viable so oil price volatility will be discussed and the impact on the future viability of bio-renewable resources considered. The propylene value chain into acrylics will be presented as an example to provide insights concerning the likely sustainability of key organic raw materials for the coatings industry.

Notes:

## 10.3 The first biobased diisocyanate and a new building block for polyurethane coatings: PDI

### Dr. Gesa Behnken, Covestro

*Co-Authors: Christoph Eggert, Andreas Hecking, Berta Vega Sánchez*

Sustainability is increasingly impacting on the purchasing decisions of customers, brand owners and consumers. Biobased polyols have existed, but up to now the limiting factor in developing biobased polyurethanes has been the crosslinkers. Bayer Material Science has solved another part of the puzzle of how to develop more sustainable polyurethanes with the release of DESMODUR® eco N 7300. This is a new solvent-free aliphatic polyisocyanurate and the first polyurethane crosslinker on the market with a significant renewable content that has not come at the expense of performance. It is based on a new diisocyanate, the pentamethylene diisocyanate, which is based on a raw material produced via a fermentation process directly from biomass and that delivers a significantly lower carbon footprint.

This new diisocyanate opens up a new technology platform and possibilities for a number of new derivatives and application fields. This presentation will highlight the latest developments based on pentamethylene diisocyanate.

Notes:

## 10.4 The next generation of highly-sustainable binders

### Mr. Simon Austin, Synthomer

*Co-Authors: Bob Foster, Dr Martin Conrads*

Over the last decade developments to make binders more sustainable have centred on reducing VOC´s and APEO values, saving energy and reducing waste in the production process, as well as ensuring longer service times for the end product. Now the next stage is underway: the use of raw materials from renewable sources as feed-stock.

However the key challenge for suppliers is to ensure that new materials don´t compromise in quality and performance. Therefore the choice of available raw materials from sustainable sources is still currently limited.

During this presentation modern polymer dispersions for the coating industry based on a monomer from renewable sources will be presented and compared with state of the art products based on petrochemical raw materials. The focus will be on low VOC paints, showing how Synthomer binders with high biobased content perform in such paint formulations.

However the key challenge for suppliers is to ensure that new materials don´t compromise in quality and performance. Therefore the choice of available raw materials from sustainable sources is still currently limited.

During this presentation modern polymer dispersions for the coating industry based on a monomer from renewable sources will be presented and compared with state of the art products based on petrochemical raw materials. The focus will be on low VOC paints, showing how Synthomer binders with high biobased content perform in such paint formulations.

Notes:

## 10.5 Lactide: Reduced alkyd resin viscosity and improved drying time in decorative coatings

### Dr. Armin Michel, Corbion

Solvent-based decorative coating formulations have changed over time due to increasingly stricter regulations on emissions defined as volatile organic compound (VOC) content of the final paint. To meet the emission targets, high-solid systems with lower VOC content have been developed, opening the door to new challenges like the performance coating properties including drying time behavior.

Alkyd resins that are suitable for high-solid paint formulations require low resin viscosity without impairing the coating performance. The incorporation of the unique bio-based building block L-Lactide into alkyd resins show a significant reduction in viscosity compared to reference resins enabling the development of high solid decorative coatings with benefits such as shorter drying time, improved hardness and excellent gloss retention whilst further reducing the environmental impact of solvent-based coatings.

L-Lactide can be incorporated into alkyd based resins using existing production processes. The resin formulation is modified by using L-Lactide to replace a combination of di-functional glycols and di-acids, while maintaining resin properties such as acid-value and molecular weight. L-Lactide is a cyclic ester that will undergo a ring opening reaction (ROP) in the presence of hydroxyl groups (initiator) and a catalyst allowing for a higher resin yield since no water is released during the ROP process step.

Air-drying coating formulations with L-Lactide modified alkyd resins result in a significant reduction in drying time with very good coating properties like hardness, gloss and yellowing resistance.

L-Lactide is an unique and very versatile bio-based building block allowing the industry to improve performance and sustainability of decorative coatings based on alkyd technology.

Notes:

## 10.6 Biobased paint for wood, the influence of the binder structure on performance

### Dr. Simon Kervyn, Coatings Research Institute

*Co-Authors: Mireille Wenkin, Aurélien Pirmez, Loic Poussard, Chistrina Gabriel, Waltraud Vorwerg*

This presentation will outlined the results obtained when substituting petrol based polymers by bio based polymers for coatings on wood. Two sources of biopolymer were evaluated: dispersion of bio based polyurethane with various chains lengths and starch extract. Those two binders were characterized to probe their visco-elastic behaviour, scratch resistance, rheological behaviour, abrasion resistance ... After the characterization, the chemical structures of the polymer were modified in order to enhance the desired properties for the film. PU dispersions were applied on wood and the properties matches or outperformed the properties of commercially available paints based on petrol based binders. The results obtained will be presented for the first time.

Notes:

## 10.7 Biobased prototypical coatings for interior metal surface application

**Dr. Matthias Wanner, Fraunhofer IPA**

*Co-Authors: Ulrich Christ, Jevgenij Lazko, Loic Poussard, Christina Gabriel, Waltraud Vorwerg*

Organic polymer materials for coatings based on fossil fuels are intended to be increasingly substituted by polymers based on renewable resources. The present study explores the potential of biobased materials for their application as organic matrices. By trend, biobased materials recently available are expected to exhibit a lower performance if used as an organic coating matrix than present fossil fuel based ones do, which have been optimised within decades. Therefore, the present study focuses on the potential use of such biobased coatings for lower performance application, i.e. interior application, here in particular for metal substrates.

We used special starch derivatives as well as polyurethane dispersions, which base on esterified fatty acid dimers. By a preselection process the most promising matrix variants were identified. In a second step these matrices were furnished with functional additives and the effects of such functionalisations were characterised regarding anticorrosive, mechanical and adhesion properties.

Notes:

# Session 11: Exterior Durability & Corrosion Resistance, Part I

## 11.1 Understanding the barrier properties of wood coatings on water transport by means of magnetic resonance

**Ms. Özlem Gezici-Koç, Tue**

*Co-Authors: S.J.F. Erich, H.P. Huinink, O.C.G. Adan and L.G.J. van der Ven*

As a result of the regulations, a transition towards waterborne coatings has been made. Generally, waterborne coatings are intrinsically more sensitive to water. This higher sensitivity to water may negatively affect the durability of both wood and coating. One of the key research question focuses on the importance of the interface on the barrier properties (both wetting and drying) of the coating. To answer this question, Magnetic Resonance Imaging on both the uncoated and the coated wood is applied. Our studies focus on understanding the moisture transport through alkyd (solvent and waterborne) and acrylic model coating formulations applied on teak, oak and pine. We will present results showing that the moisture permeability depends on the specific combination of wood and coating. Surprisingly, we found that diffusion of each coating is same and the difference in permeability is the reason of difference in solubility.

Notes:

## 11.2 Characterizing the surface structure - From substrate to topcoat

### Dr. Michael Osterhold, Consultant

Besides colour, effect and gloss, the visual impression of a painted surface is influenced especially by the surface structure (levelling, waviness, orange peel, appearance). To characterize the structure of a surface, mainly two different measuring methods - profilometry and 'wave-scan' - have been established in the automotive and paint industry. The mechanical profilometry in combination with Fourier techniques allows detailed investigations concerning the surface topography from substrate to topcoat. For example, influences of the substrate on the different coating layers and other effects can be evaluated. Basic relations and relevant application examples from the areas 'metal, plastics, coating', investigated with profilometry and 'wave-scan' in the course of time, will be summarized in this paper. Results obtained from coating surfaces with lower gloss will also be presented.

Notes:

## 11.3 Accelerated screening of new inhibitor pigment systems

### Dr. Patrick Dodds, Tata Steel

Analysis of performance of new corrosion inhibitor species has previously involved additions to commercial paint systems and screening over 1000 hours in the ASTM B117 salt spray test. The Scanning Kelvin Probe (SKP) offers rapid screening within 24 hours that allows the researcher to assess the effectiveness of inhibition of cathodic delamination by the new inhibitor species. The effect on initiation inhibition, kinetics of delamination and changes to the rate-limiting step of the corrosion cell can be obtained. After screening with the SKP and model coating, the effects of the inhibitor can be further assessed with electrochemical techniques and salt spray testing within a commercial coating. This means that within 24 hours, the researcher can eliminate any in-effective inhibitors and match the performance to a standardised system like strontium chromate. The technique has been particularly useful in the search to replace hexavalent chromate as a pigment within a commercial coating system, resulting in a rapid, structured testing regime producing a system that out performs a current chrome-free technology, in both lab testing and accelerated salt spray testing.

Notes:

## 11.4 Functional additive technologies for improved metal surface protection

**Mr. Esko Pajunen, Huntsman Pigment and Additives**

Huntsman Pigments and Additives is a business division of Huntsman Corporation. We are one of the most diversified global suppliers of pigments and additives. We offer a broad range of specialty titanium dioxide pigments, color pigments and functional additives including pure synthetic $BaSO_4$ and ZnS products. This presentation focuses on an introduction of $BaSO_4$ product technologies providing high performance in coatings for metal surfaces such as coil, automotive and protective coatings. Additionally, examples of ultrafine $BaSO_4$ to improve corrosion resistance in primers will be discussed. Our concept of functionalized barium sulphate products is also introduced. An example will be given about formation of covalent bonding between the $BaSO_4$ particle surface and the surrounding matrix resulting in enhanced corrosion resistance.

Notes:

## 11.5 Advanced waterborne barrier coatings for the corrosion protection of structural steel

**Dr. Brian Skerry, Sherwin-Williams**

*Co-Authors: S.L. Hilton, M.S. Sibbald, I. Martin-Fabiani, A. Fortini, R.P. Sear, J.L Keddie, F.D'Agosto, M. Lansalot, E. Bourgeat-Lami, J. Lesage de la Haye, M-L. Koh, B. Naden, C. Boncompain*

Corrosion protection of structural steel is often provided by solvent based barrier coatings with associated environmental concerns. This work describes progress towards a one-component waterborne barrier coating competitive in performance with two component solventborne coatings. Recent developments in polymers and organic/inorganic nano-composites allow coatings with greatly enhanced water vapour and liquid barrier properties. Here, surfactant-free latices stabilized by amphiphilic block copolymers, with self-crosslinking functionality, and which can be combined with organic/inorganic hybrid latices form functional barrier paints. Film formation conditions have been identified to cause the stratification of bimodal binder particles with a preferred type of particle at the top surface. The effect depends on the particle sizes and size ratios. The stratified structure potentially allows the tailoring of surface and bulk properties in a one-step process. These clear latices can also be formulated with pigments to provide environmentally compliant yet effective water based corrosion protective coatings.

Notes:

## 11.6 Anticorrosive performance of Zn-Mg-Al coatings in automotive multi-metal process

### Dr. Ulrich Christ, Fraunhofer IPA

*Co-Authors: Dr. Stefanie Wunder, Dr. Christina Bauder*

The anticorrosive properties of a new generation of hot-dip galvanized steel with zinc-magnesium-aluminium coatings (ZM) were investigated in combination with an automotive electro-dipping paint (EC) and compared to references of standard hot-dip galvanized coatings. In general, ZM coatings consist of three types of phases with different wt% of zinc, magnesium and aluminium.

Two types of ZM substrates were treated with a zinc phosphate and two different new thin layer pretreatments after which an electro-dipping paint (EC) was applied. Surface analytical studies by SEM indicate that the pretreatment layers cover the surfaces throughout. The new thin layer pretreatments are only a few 100 nm thick and are mainly composed of zirconium oxide, embedded in a thin organic layer. In comparison, the zinc phosphate layer has a thickness of 1-2 µm, depending on the orientation of the zinc phosphate crystals.

In various corrosion tests, such as cyclic tests with a high corrosive load e.g. the old VDA (VDA 621-415) as well as intensified outdoor weathering (VDA 621-414), the new pretreated and EC coated ZM substrates showed a consistently lower undercreepage at scribe and thus a higher corrosion protection as the standard zinc coatings Z and ZE. Under the conditions in the new VDA/VDEh cyclic test (VDA 233-102, VDEh SEP 1850) both, the standard zinc coatings and ZM materials showed a very low undercreepage which complies with a high corrosion protection, accordingly in case of the ZM-materials, compared to Z and ZE, only a slightly lower undercreepage occurred. Furthermore, round robin tests showed a good reproducibility of the tests and the corrosion behavior. The ZM substrates showed significantly lower edge corrosion than the references Z and ZM.

Electrochemical studies underpin the good anticorrosive performance of the ZM substrates in combination with pretreatment and EC. SKP investigations show a very homogeneous surface of the pretreated ZM substrates with both zinc phosphate and thin layer pretreatments. High barrier properties of EC-coated systems are found with electrochemical impedance spectroscopy.

These results suggest that the ZM materials have a great potential for use in the OEM sector.

We thank FOSTA Stiftung Stahlanwendungsforschung for funding and the project consortium companies for supporting this project.

Notes:

## 11.7 Influence of SiO2 on the corrosion performance of carbon steel

### Ms. Laura Vivar Mora, NSIRC / University of Leeds

Co-Authors: Professor Anne Neville, Dr Richard Barker, Dr Sanjeev Naik, Aurelie Rexach

Corrosion has a significant impact both economically and environmentally on almost all the world's metallic infrastructure and can compromise its integrity if no measures are taken to prevent or control it. Coatings are typically used to provide a barrier between the metal substrate and its environment to avoid corrosion.

This research is focused on understanding how the addition of nanoparticles to a coating formulation can improve the corrosion protection of carbon steel. Specifically, in this study silica and cerium oxide nanoparticles are added to polysiloxane-based matrices to provide improved barrier and hence better corrosion protection of carbon steel. Corrosion protection and coating durability are investigated using different methods such as electrochemical impedance spectroscopy (EIS), polarisation curves and salt spray.

Many challenges remain in the processing of nanostructured coating materials, and preliminary results on the influence of nano-additives on properties and performance of the coatings will be presented.

Acknowledgements:
One of the authors, L. Vivar Mora, would like to thank the Lloyd's Register Foundation for sponsoring this PhD research and would like to specially thank Richard Dawson of Lloyd's Register EMEA for his support.

Notes:

## 12.1 Encapsulated biocide ensuring low biocide content in anti-fouling technology

### Ms. Eva Wallström, EnCoat

A silica aerogel encapsulated biocide, eg. zinc pyrithione, can be used in an anti-fouling paint to control the biocide leaching from the paint. Yacht paint formulations with approximately 70 weight-% reduction of biocide, and without pigments/fillers as cuprous oxide and zinc oxide, that may cause long-term adverse effects in the aquatic environment, have been formulated and evaluated. A reduction in cuprous oxide will reduce the raw material cost of an anti-fouling paint, as cuprous oxide currently is one of the most expensive raw materials in an anti-fouling paint.

The test paint formulation has approximately the same technical properties as a commercial solvent-based anti-fouling paint. The gel particles are hydrolysed/worn down throughout the polishing process thus giving a constant effective release of biocide in the coated surface layer.

The test paint has during a summer season proven similar anti-fouling efficacy on rafts compared to a commercial product and to show the same effectivity against fouling on yachts in sea water. Technical properties of the gel as a raw material, the wet paint and the anti-fouling paint film are discussed and examples of practical experiences are given. The gel technology, can also be used with other biocides more suitable for large vessels or for other coating applications

Acknowledgements:
For financial support: The Danish Environmental Protection Agency (DEPA, incl. representatives from the DEPA and industry following the project)
As cooperation partners:
Kjeld Schaumburg, Roskilde University (Centre for Interdisciplinary Studies of Molecular Interactions)/CISMI/FiBaC and
Björn Sivik, Bjarga AB (Both with regard to knowledge on aerogel technology and producing samples on demand)
Jesper Højenvang, Danish Sailing Association (Raft tests in Jyllinge and Horsens harbours etc.)
Professor Sergey Dobretsov, Sultan Qaboo's University, Oman (Giving us opportunity to test in Oman)
Annika Vaksmaa (Raft tests and fouling reported from Muscat, Oman)
Agnete Krabbe Katholm (Raft tests and fouling reported from Sydney, Australia)

Notes:

## 12.2 Intrinsic self-healing coatings

### Ms. Anastassija Wittmer, Fraunhofer IFAM

*Co-Authors: A. Brinkmann, Dr. C. Schreiner, Dr. V. Stenzel, Prof. Dr. A. Hartwig*

Cracks on microscopic scales caused by mechanical stress or thermal fatigue can lead to material failure and indispensability. Self-healing polymers feature the ability to heal after damage to maintain and prolong the lifetime of the material.
Inspired by nature, a huge number of self-healing concepts were proposed in recent years.(1) The self-healing concepts can be divided into two categories: (i) extrinsic, which offers self-repairing through a self-healing agent (e.g. microencapsulated) and (ii) intrinsic, which can realize self-repair by a special molecular architecture with reversible covalent or non-covalent bonds. The latter concept describes a supramolecular polymeric system that interacts through weak physical bonds like hydrogen bonds. Those molecules are able to build supramolecular networks and show polymer-like behaviour in their rheological and mechanical properties.(2) (3) (4) (5) (6)

After crack or scratch formation, the supramolecular polymeric network is able to self-heal through the reversible breaking and reforming of non-covalent bonds with an external stimulus (heat, pressure, moisture or UV-irradiation) or autonomously. The image below shows the self-healing of cracks formed during Erichsen cupping in a PUR-clearcoat on an aluminium plate. The self-healing procedure takes nine minutes and was carried out on a heating plate at 110 °C.

The intrinsic self-healing mechanism via multiple hydrogen bonds is a dynamic process which is influenced by external factors like temperature, humidity and others. These influences define the possible process window for technical applications of such coatings.

Notes:

## 12.3 Intrinsic healing protective coatings reinforced with hybrid nanocontainers

### Ms. Evangelia Karaxi, National Technical University of Athens

*Co-Authors: I.A. Kartsonakis, E.P. Koumoulos, C.A. Charitidis**

This study focuses on the design, development and validation of a new type intrinsic healing coatings for corrosion protection of metal alloys. For this purpose, ceria ($CeO_2$) and cuprous oxide ($Cu_2O$) nanocontainers modified with a polymer shell synthesized via ARGET-ATRP comprising of a low $T_g$ polymer (poly(n-butyl acrylate)) were dispersed in epoxy resin. A three-layer coating was applied on HDG steel via a roller applicator. The expected result was the production of a coating combining the durability of an epoxy coating with the reflow and deformation ability of a low $T_g$ polymer and the mechanical stability and robustness of inorganic particles acting as reinforcing elements within the coating matrix. The aforementioned composite materials were characterized through TEM, FTIR and XRD. The applied coatings were studied through FE-SEM cross-section images and EDS elemental mapping.
Corrosion studies via EIS revealed that the coatings have enhanced barrier properties. Moreover, studies on artificial defected coatings via EIS and LEIS after the formation of an artificial defect, demonstrated an increase of polarization resistance and a decrease of the anodic current densities as the immersion time elapsed, suggesting possible self-healing effect.

Acknowledgements
This work was supported by the project: «**S**ELF-**HE**ALING NANOMATERIALS FOR PROTECTION OF METAL A**LL**OYS», CODE 3456 & ACRONYM: "SHELL", for the National Action "ARISTEIA II" of the Operational Programme "Education and Lifelong Learning", by the European Union - European Social Fund (ESF) and National Funds through the Operational Program "Education and Lifelong Learning" (ELL).

Notes:

## 12.4 Advanced functional coatings – Adding value to steel

### Dr. Digvijay Thakur, Tata Steel

Organic coated steel products have been fulfilling the demand by customers for superior performance for longer service life (e.g. using anti-corrosion protective organic coating) over past several decades. With the advent of functional materials and nanomaterials, an enormous opportunity is presented to produce coated steel products with variety of advanced functionalities beyond corrosion protection alone. Accordingly, novel functional materials based coatings on steel can potentially benefit various market sectors including but not limited to building envelope, automotive etc. The talk will provide an overview of the current efforts underway at Tata Steel R&D to generate differentiated steel products incorporating such advanced functional coatings on steel.

Notes:

## 12.5 The use of easy-to-clean, antimicrobial coatings in the food industry

**Dr. Heidi Van den Rul, Sirris Smart Coating Application Lab**

*Co-Authors: Daan Dewulf, Joey Bosmans*

Easy-to-clean and antimicrobial coatings can be a strategy to facilitate the cleaning and disinfection of production equipment in the food industry. They can reduce the number of cleaning cycles and cleaning products and the adhesion of bacteria and biofilms. Easy-to-clean coatings are used in diverse applications where fouling needs to be avoided. For food producing companies, special attention needs to be paid to the food safety of these coatings. But also aspects like overdosing of biocides, durability of the coating, re-applying on the coatings on existing equipment etc. play an important role in the implementation of these coatings in the food industry. In this presentation an overview and benchmarking analysis study is shown of existing commercial and innovative (e.g. superhydrophobic) easy-to-clean coatings that can be used in contact with food.

Notes:

## 12.6 Hydrophobation of timber with organosilicon compounds – an update from the SILEX Life+ project

**Dr. Jean Paul Lecomte, Dow Corning**

*Co-Authors: Joris Van Acker, Magdalena Kutnik, Jan Van den Bulcke, Mathilde Montibus, , Lise Malassenet, Sabrina Salvati, Sarah Derocker*

The SILEX project "Improving sustainability of construction materials using innovative Silicon based treatment" is a Life+ project with reference LIFE+11 ENV/BE/1046.

This project intends to demonstrate that a new class of compounds can be used for wood treatment for an extended durability and that the testing phase could be accelerated according to new testing methodology. The project aims at demonstrating that treatments with silicon-based hydrophobers have a lower impact on the environment with a lower need of biocide use than the current techniques.

Various silicon-based hydrophobers (non-film-forming) were screened and tested as "impregnation hydrophobers" on three European wood species. Different application methods were used as well (vacuum impregnation, brushing).

This update of the project intends to present some preliminary results showing potential to use this technology.

Notes:

## 12.7 Varnish development for enhanced gas barrier and antimicrobial properties of food packaging

### Dr. Benjamin Naden, PRA

*Co-Authors: Laura Pilon, Marit Kvalvåg Pettersen, Lars Axelsson, Lorena Rodriguez Garrido*

This paper describes the development of a varnish with antimicrobial, oxygen and water barrier properties to be used in the food industry. The aim is to reduce reliance on metallised, multi-layered films for high barrier applications (which may be complex to manufacture and difficult to recycle) with a single varnish formulation that can be easily applied to a variety of substrates, allowing shelf-life extension of foods and reducing food waste.

Materials were selected based upon safety in use and for the performance properties of the varnish when applied to the film substrate. Gas barrier performance was realised by the tortuous path to gas diffusion provided by high aspect ratio minerals, optimum exfoliation of which was achieved by a process of in-situ polymerisation, requiring clay modifiers with high thermal stability. Antimicrobial activity was achieved by incorporation of volatile essential oils. Property improvements were observed by OTR, WVTR, and antimicrobial challenge testing.

Notes:

# Session 13: Packaging Coatings

## 13.1 Functional coatings for biodegradable food packaging

### Dr. Sabine Amberg-Schwab, Fraunhofer ISC

*Co-Author: Dr. Daniela Collin*

Today, fresh food as well as convenience food is sold in packages. Hygienic conditions, long shelf life and easy availability of the packed products account for our standard of living. However, this comfort contributes to environmental pollution in a significant way, as packaging is mainly achieved by plastic materials. Pollution by plastics is becoming extremely evident in our oceans and even in ground water and poses a threat to flora and fauna and human health, too. Biodegradable packaging materials have been on the market for many decades. However, there are severe restrictions preventing their broad application in food packaging, as these materials do not possess the requested properties, such as a sufficient barrier against water vapor, oxygen or flavors.

The idea has been to improve their properties by a biodegradable coating. At Fraunhofer ISC, high-barrier coatings based on a class of materials with glass-like structural units, the so called ORMOCER®s have been developed. As these coatings are originally not biodegradable, the ORMOCER®s had to be adjusted to become biodegradable and at the same time preserve their barrier and functional properties. These novel biodegradable coating materials can play an important role in the future design and development of innovative and more eco-friendly packaging materials.

Notes:

## 13.2 Effect of the composition of new bisphenol A non-intent coatings for interior can coating

**Dr. Katharina Weber, Fraunhofer IPA**
*Co-Authors: Volker Wegmann, Matthias Wanner, Michael Hilt*

Consumers expect canned foods to be free of toxicologically critical substances. Since one of the starting materials for traditional interior can coatings, bisphenol A (BPA), is currently an issue of a public debate due to its estrogenic effects, alternatives have to be developed. Additionally, the upcoming European REACH regulation induces Cr-free post-treatment processes in the tinplate production. This requires the development of new, reliable and well-understood solutions.

Recently, promising polyester-based coatings have been successfully brought to the market by several suppliers. So far, there is little information about the effect of the polyester-based coating formulation on the functional properties of the coating. We report a systematical study on polyester-based coatings for the insides of tins regarding their overall performance with varying composition. Typical requirements for can coatings, such as sterilization resistance, flexibility, corrosion protection and suitability for Cr-free tinplate substrates, are of particular interest.

Acknowledgement
We thank Annette Krug for the preparation of most of the investigated samples, for the mechanical tests and some of the electrochemical experiments. Dr. Jochen Hollaender from Fraunhofer IVV is gratefully acknowledged for his input and expertise in EIS measurements. We thank Lukas Aktas, who has performed some of the electrochemical experiments. We thank Rainer Bürsner and Markus Keuerleber for supportive work. We thank Prof. Dr. Christian Mayer and his team for the solid state NMR experiments. Dr. Angela Störmer and Maria Hoppe from Fraunhofer IVV are acknowledged for a fruitful collaboration within this AiF project. We thank BMWi (Federal Ministry for Economic Affairs and Energy in Germany) for financial support and our industry partners for providing us with sample material and for sharing their experience with us.

Notes:

## Session 14: Exterior Durability & Corrosion Resistance, Part II

## 14.1 Green inhibitors for corrosion protection of metals,

**Dr. Anna Szelag, Tata Steel**
*Co-Author: Dr. Sohail Hajatdoodst*

Corrosion inhibitors have gone through many stages of evolution. Despite steady progress, there is still need for improvements. Serious environmental and health concerns about the chromating process have generated a great deal of interest in developing chrome-free conversion coatings. That is why the scientific community began searching for chrome free, environmentally friendly "green corrosion inhibitors". Recent innovations in the area of organic and inorganic corrosion inhibitors offer new opportunities to develop novel compounds for metal protection.

This presentation will focus on different types of corrosion inhibitors (CIs), their mechanism (cathodic or anodic process) and anticorrosion properties. Ecological alternatives in corrosion protection systems, which can be used in the future, will be presented. Developments described in the lecture mainly will concern the application of polymeric materials, plant extracts, ionic liquids etc. as corrosion inhibitors for metal surfaces in different media.

Recent papers and patents in the field of corrosion inhibition will be presented.

Notes:

## 14.2 Organic-inorganic hybrid aerogels as coating materials

**Ms. Katarzyna Suwala, Industrial Chemistry Research Institute**

Aerogels are a unique class of microporous materials with very low density. Typical aerogel consists of 95 - 98% air while the remaining is a thin, delicate structure of the material, e.g. silica, coal etc. Aerogels are characterized by very high surface area and low thermal conductivity. Due to these properties aerogels are used in many applications such as thermal insulation, acoustic, catalysts, thin film dielectrics etc. Despite the large research activity in this area and the time elapsed since obtaining the first samples of aerogel only silica aerogel has been implemented into industrial production. It is expected that the hybrid aerogels are cost-effective, environmentally friendly and energy efficient alternative to the traditional insulation. The effect of organic component will be presented based on the results of supermolecular structure examination, thermal conductivity and porosity as well as dust release evaluation.

Notes:

# Session 15: Functional Coatings, Part II

## 15.1 Assessment of advanced easy-clean durable coatings

**Mrs. Anna Wojdyla-Cieslak, NSIRC**
*Co-Authors: Alan Taylor, Ian Boyd*

Highly repellent surfaces (anti-fouling, anti-icing) are of considerable potential for research with benefit to the transportation, construction and power generation sectors. Conventional, highly repellent coatings have only limited industrial capabilities, due to the low mechanical durability, complex preparation, and awareness of the availability of the coatings with all required properties. New advanced coatings systems having inorganic additives might have the potential to overcome the limitations of existing technologies, by linking together high repellency and mechanical robustness. Nevertheless, the lack of understanding wear mechanism in such coatings and lack of comparative test methodology that compares the durability of highly repellent surfaces hamper further development of novel coatings.

This study introduces a new approach to testing that enables highly repellent characteristics to be examined against the degree of mechanical damage. The new assessment criteria indicate that by addition of novel inorganic building blocks improved performance of the coating can be achieved.

Notes:

## 15.2 Benchmarking dirt pickup resistance of commercially available architectural exterior coatings

**Mrs. Amy Cooper, Chemours**

Architectural paints designed to resist the accumulation of dirt have a positive impact on the durability and visual appeal of exterior coatings. This presentation summarizes learnings to date from an ongoing evaluation of dirt pickup resistance in commercially available exterior architectural coatings from 11 countries around the world. This includes a discussion of accelerated versus natural testing; regional differences, challenges, and opportunities.

Notes:

**Additional Notes**

## Poster 1: Anti-microbial polyethylene surfaces

### Dr. Konstantin Siegmann, North Dakota State University

*Co-Authors: Fernanda Rossetti, Joachim Köser and Martin Winkler*

Any surface that is in continuous contact with water will eventually be covered by "biofilm", that is, microorganisms will adhere to it. For many surfaces, like ship hulls, medical devices and sanitary installations, biofilm is unwanted and even problematic or dangerous. We want to permanently prevent biofilm formation on plastic surfaces such as polyethylene by constructing a non-leaching biocidal surface.

It is well known that certain quaternary ammonium compounds kill bacteria. Our approach is to compound quaternary ammonium salts (quats) into polyethylene. Because they are not miscible with the non-polar polyethylene, quaternary ammonium cations will segregate to the surface and form a biocidal layer. Anchoring of the quats in the matrix is achieved by a long hydrocarbon tail on the quat which is miscible with the polyethylene and therefore compatible with the matrix.

First results show that our approach is valid. The biocidal activity of the treated polyethylene surfaces was demonstrated by populating them with S. aureus or E. coli, followed by fluorescent "dead/live staining" and microscopical analysis. It is found that there are more death bacteria on the treated surface than on the surface without quat. Further, it was shown that selected quats do not leach into water (or only do so initially) and that the biocidal activity is retained after storing the specimens in hot water for one month.

Notes:

## Poster 2: Self-crosslinkable latex with covalently linked flame retardent

### Mrs. Adéla Rückerová, Eindhoven University of Technology

*Co-Authors: Jana Machotová, Lucie Zárybnická, Miroslav Večeřa*

In this study the synthesis of a novel flame retardant based on halogenophosphazene derivative and its application in waterborne coatings based on self-crosslinking latexes was investigated. Hexaallylamino-*cyclo*-triphosphazene was synthesized by nucleophilic substitution of hexachloro-*cyclo*-triphosphazene with allyl amine. Latexes of functionalized core-shell particles bearing in the structure hexaallylamino-*cyclo*-triphosphazene molecules were prepared by the semi-continuous non-seeded emulsion polymerization of methyl methacrylate, butyl acrylate and methacrylic acid as main monomers. For interfacial crosslinking, diacetone acrylamide was copolymerized into the shell layer of latex particles to provide sites for subsequent reaction with adipic acid dihydrazide. Because of the double bonds present in hexaallylamino-*cyclo*-triphosphazene molecule, this kind of flame retardant can be incorporated into the structure of emulsion copolymers by radical copolymerization. The increasing concentration of hexaallylamino-*cyclo*-triphosphazene was found to have the positive effect on the flame stability of coating systems and also increased the crosslinking density of latex particles. The incorporation hexaallylamino-*cyclo*-triphosphazene did not affect transparency, flexibility, toughness and adhesive properties of resulting coatings. Moreover, the presence of the novel flame retardant decreased water sensitivity and increased the flame stability of coatings in terms of reduced total heat release, decreased amount of released smoke and drop in maximum average rate of heat emission which indicated a slower flame spread during the material combustion.

Notes:

## Poster 3: Molar mass of acrylic microgels in relation to coating properties of self-crosslinking latexes

**Dr. Jana Machotová, University of Pardubice**
*Co-Authors: Štěpán Podzimek, Hedvika Zgoni, Jaromír Šňupárek*

The object of our interest focused on self-crosslinking latexes based on keto-hydrazide crosslinking system. The latexes of core-shell particles were synthesized by emulsion polymerization of methyl methacrylate, butyl acrylate and diacetone acrylamide as main monomers. The structured latex particles consisted of an internally crosslinked core and a shell layer containing DAAM repeat units within the polymer chains. Molar mass of the shell copolymers was gradually reduced with different amounts of isooctyl 3-mercaptopropionate included in the synthesis of the shell layers and was determined using the conventional SEC-MALS technique and the novel A4F-MALS method. Whereas SEC was found to be convenient for characterization of low molar mass shell copolymers without crosslinked structure, A4F was proved as a very efficient technique for characterization of high molar mass copolymers and core-shell microgels.

Furthermore, the attention was focused on the optimization possibility of desired coating properties of self-crosslinking latexes by varying the molar mass of the shell layer of latex microgel particles. The results demonstrated that decreasing the molar mass of shell copolymers resulted in a drop in the minimum film-forming temperature of the latexes, obviously due to plasticization effect of low molar mass copolymer molecules leading to enhanced coalescence at lower temperatures. It was demonstrated further that with decreasing the molar mass of the shell layer of latex particles, the initial speed of water absorption into the latex film was reduced, while the total water uptake after a long-term immersion was increased as a consequence of better coalescence of latex particles, which was caused by favored mutual inter-particle diffusion of low molar mass shell copolymers.

Acknowledgements
The Technological Agency of the Czech Republic (TA02000011) is gratefully acknowledged for supporting this work.

Notes:

## Poster 4: Not only paints! Electrochemistry can also protect and colorize

**Dr. Artur Maciej, University of Pardubice**

Anodic oxidation of is a technique of metals and alloys surface treatment which allows to form thin conversion (usually oxide) coatings. The metallic materials which are commonly treated in such a way are: titanium, aluminium, magnesium, niobium, tantalum, zirconium and their alloys. The oxide coating is responsible for corrosion resistance improvement. Some of the metals may also be colorized during the anodic treatment. The parameters of anodizing, e.g. type of electrolyte, voltage influence the resulting color and its intensity. The colors are most often associated with the passive layer thickness, which is proportional to the applied potential. There is the chromatic scale which shows the colors feasible to obtaining, listed in sequence of growing potential: yellow – purple – blue – light blue – silver – yellow – pink – violet – cobalt blue – green –yellow/green – pink – green.

The results of investigations on colorizing of galvanic zinc alloys are presented in this work.

Notes:

## Poster 5: Thermal stability of heat resistant pigments

**Mr. Tamás Pozman, Festékipari Kutató**
*Co-Authors: Lívia Molnar-Nagy; Anikó Bende*

We examined how the colour of various heat resistant pigments changes thermally in different binder matrix (non-thermoplastic organic resins, polysiloxanes, silicates and enamels).

The heat resistance of pigments refers to the maximum temperature which a pigment can tolerate. According to the test method  E is calculated from colourimetric parameters of pigmented coatings before and after heat exposure measured on the cooled back surfaces at RT. This method ignores the reversible thermal colour shift, which may remain unobserved. The colour of coating for stoves, fireplaces, smoke pipes etc. should be independent from the actual surface temperature which can vary up to several hundred degrees from bottom to top during operation, consequently reversible colour change has to be avoided as far as possible.

In our study we monitored the reversible and irreversible colour change of several heat resistant inorganic pigments in various coating systems during the heating up process.

Notes:

## Poster 6: Exploration of cerium-doped inorganic pigments as a concept for UV-protection of coatings

**Mrs. Monika Eva Bohem, Fraunhofer IPA**
*Co-Authors: Marc Entenmann, Thomas Schleid*

Providing UV-protection is one of the pivotal fields of interest in coating systems and painting technology. UV-protective inorganic pigments represent a promising area of materials research to overcome the problem of photodegradation and to increase the lifetime of the coating as well as of the substrate beneath. In this work, we synthesized Ce3+ -doped La2[Si2O7] as a phase-pure product via solid-state reaction. The characterization of the system revealed relevant photoluminescence and scintillation properties. Furthermore, the spectral patterns of the initial synthesis product could be maintained after particle size reduction and were also detectable after incorporation into a coating formulation. We could show that the pigment particles outreach the functional aspect of absorption, by combining absorption properties and fluorescence as a form of downconverted re-emission of higher energetic radiation. The potential of an innovative UV-protection concept for coatings, with regards to stability and environmental friendliness, could be successfully pointed out.

Notes:

## Poster 7: Improvement of corrosion protection of offshore wind turbines by mechanical characterization of protective coatings

**Mrs. Isbelis Lopez, Northumbria University**
*Co-Authors: Dr. P.S Leung, Dr. Fawad Inam and Dr. Clare Till*

The wind energy industry is increasingly moving to offshore areas and increasing the wind turbines' size and number. Since these are unmanned structures with limited access, costs of repairs are significantly higher than for onshore repairs and higher reliability of the corrosion protection system is required. In recent years research has directed interest towards developing coating materials capable of tolerating hostile marine climates with temperatures ranging from -25°C to +25°C with minimum maintenance [1]. The corrosion protection of metals has long been achieved by the use of organic coatings, such as epoxy resins, which act as a physical barrier by tightly adhering to the metal against corroding agents. When epoxy resins are applied onto metallic substrates, intrinsic residual stresses originate within the coating due to its adhesion to the substrate and the prevention of the natural movement of the coating. Residual stresses have an adverse effect on adhesion and may produce coating failure [2, 3]. Three epoxy-based paint systems, used on towers and transition pieces, were airless-sprayed onto stainless steel shims and post-cured to produce single layer coatings. The samples were analyzed through optical scanning to study their surface characteristics and profile parameters. Flexural tests on simply supported beams in the three point bending mode were performed using dynamic mechanical analysis. A mathematical model using composite beam stress analysis was developed to determine the Young's modulus of the coatings. This knowledge can give an understanding of coating properties that could lead to further improvements in the coating and corrosion protection of offshore structures in general. A mathematical model will be developed which will predict the mechanical behaviour of coatings exposed to marine conditions, and the effect of internal stresses during their lifetime.

Notes:

## Poster 8: Evaluation of the physical-mechanical and anticorrosion properties of paint films using nanotechnology

**Mrs. Katerina Nechvílová, University of Pardubice**
*Co-Author: Andréa Kalendová*

This paper is focused on paint films, which is examined by using nanotechnique. The paint film consists at least three components such as binders, pigments of fillers and solvent. The properties of paint films, which is based on epoxy-ester resin was measured by nanoscratch technique. Organic coatings were prepared in the following volume concentration of selected pigment PVC = 0.5 and 3 %. Non-pigmented epoxy-ester resin was used as a blank sample to compare. The pigments particles were selected at nanoscale for better study using nanotechnique. The other binder will be selected to opposite of epoxy-ester resin to the new one will be more hardness. These systems will be exposed to the corrosive environment. Subsequently, it will be comparative according elasticity, Young´s modulus and force to be applied to deform.

Notes:

## Poster 9: Anticorrosion efficiency of zinc-filled coatings containing conductive

**Dr. Andréa Kalendová, University of Pardubice**
*Co-Author: Miroslav Kohl*

The objective of this paper concerns reducing the zinc metal content in organic coatings while preserving their high anticorrosive efficiency. The two goals can be achieved by using conductive polymers as components of the protective coating. Conductive polymers are attracting interest in many branches of science and technology, including the paint industry. Among important conductive polymers are polyaniline, polypyrrole and polyphenylenediamine. Major advantages of such conductive polymers include easy preparation, high stability and non-toxicity, owing to which such substances can be used as ingredients in organic coatings. Polyaniline phosphate, polypyrrole phosphate and polyphenylenediamine phosphate were investigated as the conductive polymer. These pigments were specified using physico-chemical parameters. Organic coatings were prepared by combining zinc metal with these pigments. The mechanical resistance of the prepared films was evaluated using the results of mechanical tests. The anticorrosive efficiency of the prepared films was evaluated using the results of direct corrosion tests.

Notes:

## Poster 10: Unsaturated Polyester microstructures on steel: Mapping coatings on a nanometric scale by AFM

**Ms. Sara Muhler, Winn & Coales (Denso)**
*Co-Authors: J.A. Wharton, J. Zekonyte*

This work aims to elucidate the relationship between the morphology of unsaturated polyester coatings and the corrosion performance using Atomic Force Microscopy and Kelvin Probe Force Microscopy. Once this relationship is known, formulators and applicators will be able to identify the ideal structures to maximize performance and reduce corrosion.

Notes:

## Poster 11: Synthesis of chemical resistant polymers through formation of cyclic carbonates

### Ms. Annett Halhuber, Niederrhein University of Applied Sciences

*Co-Authors: Katrin Mathea, Bernd Strehmel*

Epoxidized synthetic and renewable raw materials react with carbon dioxide which results in the formation of cyclic carbonates. The latter possess the capability to react with amines resulting in urethanes bearing hydroxyl groups. These reactions depict important alternatives compared to the conventional isocyanate chemistry. The reaction between the cyclic carbonate and the amine proceeds relatively slow. Therefore Different catalysts were tested to decrease the reaction time, which can cover several hours without any catalyst. Nevertheless, catalysts such as triazabicyclodecene (TBD), and lithium triflate (LiOTf) result in a significant decrease of the reaction time. Further acidic and basic catalysts were included in the investigations. The cyclic carbonates were also combined with organofunctional trialkoxysilanes to enhance the mechanical and chemical resistivity of the resulting hydroxy functional urethanes. Through the incorporation of these alkoxysilanes it is possible to build up an additional inorganic network.

Acknowledgement
We thank the BMBF for financial support within the program FHProfUnt (grant no. 03FH022PX4).

Notes:

## Poster 12: A method to get round the carbonation of amines in epoxy coatings

### Ms. Marine Blain, Juxta

Epoxy resins are widely used in adhesives, sealants, filling materials, and coatings. They display good chemical resistance, high mechanical strength, and they are easy to process. There are only few examples of transparent room temperature epoxy coatings described in the literature. The main reason is the low reactivity of the epoxy + amine system at ambient temperature. To overcome this drawback, amines displaying a high reactivity must be used. However, these amines tend to be carbonated by CO2 in open air, leading to defects in the resins called blush or bloom.

Herein, we describe a way to avoid the carbonation by the synthesis of aminotelechelic prepolymers. In this study, carbonation of amine and of prepolymers was investigated by infrared spectroscopy. Then a range of epoxy materials presenting various Tg were synthesized from prepolymers and compared to polyurethane references generally used for transparent coating application.

Notes:

## Poster 13: Effect of Graphene Oxide on Self-Stratification of Epoxy-Silicone Coatings

**Mr. Shahed Ekbatani, SRBIAU**
*Co-Authors: Jamal Taheran and Dr. Ali Abbasian*

Self-stratification in last decade, because of application and its advantages and environmentally effects became important. It would be very attractive if such multi-coat systems could be applied in a single step, denoting that the applied coating stratifies spontaneously during film formation into the appropriate functional layers.

In this investigation the epoxy and silicone resins with two types of solvents were mixed and applied to glass and aluminum substrates. Graphene oxide was dispersed in epoxy resins.

The three different percents of grapheme oxide ( 0.2%, 0.5%,1%) were added to the system and based on the data, It was concluded that presence of 0.5% of grapheme oxide did not negative effect on stratification of system and also the anti-corrosion and mechanical properties compared with the system without pigment were increased.

Notes:

## Poster 14: Influence of paint quality on the environmental footprint of architectural paints

**Dr. Johan Rommen, Chemours**

Architectural paints play an important role in the decoration of interior walls. In that respect, hiding is a key performance factor, ideally without compromising other paint properties. It is generally accepted that this should be done with minimal impact for the environment. The environmental footprint can be minimized by optimizing the efficiency of paint ingredients, but also by increasing the lifetime of the paint on the wall. All these aspects will be largely influenced by the quality of the paint and its ingredients. High quality ingredients, like titanium dioxide, will influence several of these characteristics. This paper discusses how specially designed titanium dioxide can reduce the paint consumption and hence influence in a positive way the environmental footprint of the paint throughout the value chain.

Notes:

## Poster 15: Influence of titanium dioxide surface treatment on the properties of powder coatings

### Mrs. Daria Kotova, Yaroslavl Powder Coatings Plant

Powder coatings used in industry are pigmented systems usually. It is known that the introduction of pigments in the polymer matrix has a significant influence on the formation of the polymer body. Practically, all inorganic pigments are subjected to surface treatment, and many of the technological properties pigments modified, including deliberately varies the ratio of acidic and basic centers on the mosaic surface of their particles. The acid-base balance of the pigment surface will inevitably influence on the formation and properties of the polymer coating. The influence of the acid-base properties of the various chemical properties of titanium dioxide surface on the forming of the powder epoxy-polyester coatings was investigated, as the depth of crosslinking and density of the polymer network. Acid-basic-properties of titanium dioxide surface were determined using non-water heterogeneous potentiometry. The research was conducted using four grades of titanium dioxide with different surface treatment. To study were selected matte and glossy epoxy-polyester powder coatings. Polymer components in hybrid powder coating is a mixture of binders: saturated carboxylated polyester resin and a solid epoxy oligomer, which is the reaction product between liquid epoxy resin and bisphenol-A. As the matting agent used 4,5-digydro-2-phenil-1N-imideazol-1,2,4,5-benzentetracarboxylate. The rate of interaction of the carboxyl groups with the epoxy crosslinking agent is higher than the same group of polyester, since the catalytic action of the imidazole ring having a basic nature and bound in the form of salts with pyromellitic acid. Furthermore, reacting an acid with the oligomer higher due to a greater mobility than the units of the high molecular compound. Matting is realized due to the rapid formation of cross-linked epoxy resin is incompatible with epoxy-polyester, the formation of which occurs later. A comparison of the data on the gel fraction content and the acidic and basic centers content on the titanium dioxide surface shows that the greatest degree of conversion is in the case of a significant excess of the basic centers content, or when a significant excess of acid centers.

Notes:

## Poster 16: Characterization of Carbon-based Materials with Combined Raman Spectroscopy & Triboindentation

### Dr. Ian Hayward, Renishaw

*Co-Authors: Pierre Negri, Jaroslav Lukes, Jared Risan, Virginia Fergusson*

We demonstrate the benefits of integrating a Raman spectrometer with a nanoindenter to produce a valuable tool for the study of carbon-based materials, with many advantages over two separate instruments. This integration allows rapid characterization of the chemical properties of carbon – and other materials - in situ following nanomechanical testing. We demonstrate the utility of combining these techniques for the examination of amorphous Si carbide (SiC), diamond-like carbon (DLC) and highly oriented pyrolytic graphite (HOPG) samples. The analysis of the uniformity, strain, stress and defects of these materials over large areas demonstrates the variations in material's physical and mechanical properties. These results show the direct correlation between the nanomechanical properties (hardness and modulus, extracted from the nanoindenter) and the chemical information (molecular orientation, stress and conformation, provided by the Raman spectrometer). Researchers are using this important information to better understand growth mechanisms and improve growth techniques for these carbon materials.

Notes:

## Poster 17: Effect of wollastonite extender on the properties of exterior acrylic paints

### Dr. Sanem Senler Ozbil, Betek Boya ve Kimya Sanayi

*Co-Authors: İnci Yaprak Emek, Mesut Eren, Salih Zeki Kurt,*

Performance of an exterior paint film is seriously affected by its water permeability and water vapor transmission (WVT) property. Commercial acrylic binders provide adequate protection for the former, but cannot accomplish top level, for the latter. The "breathability" (water vapor and $CO_2$ permeability) of the paint is inhibited or facilitated by the binders and pigments. Also, extenders are additionally employed in order to reduce the cost, modify the volume to optimize the composition and achieve or improve technical properties. Recently we carried out an extensive investigation on the effect of partial or total substitution of the conventional calcite extenders with various mineral based extenders for opaque and semi-transparent acrylic paints. We determined that when mainly acicular shaped wollastonite (calcium metasilicate) particles of 4 and 12 µm size are employed, a profound decrease of the WVT $S_d$ values could be obtained[1]. For example, for one of the high PVC opaque paints $S_d$ (m) decreased from 0.62 to 0.07, attaining "top level" grade, when all of calcite was replaced with a mixture of small and large sized wollastonite extender. SEM micrographs confirmed the micropores which enhanced WVT. Furthermore, a weathering test of 42 months revealed that the substitution of calcite with wollastonite not only maintained the color against deterioration, but also mold growth was inhibited in the presence of wollastonite. On the other hand, in case of paints with low PVC, since binder filled most of the micropores, the effect on $S_d$ was moderate.
Since better results were obtained with larger size wollastonite extender, further work which is reported here was carried out with only low PVC paints, employing two other wollastonites having a higher ratio of acicular shape, as well as larger sizes of 28 µm and 55 µm. In contrast with the high expectations, some of the paints had rheological problems and no homogenous films could be prepared from their dispersions. Others were more successful and decrease of $S_d$ from 0.39 to 0.20 and 0.28 was obtained with 28 µm and 55 µm wollastonite extenders, respectively. SEM micrographs displayed the orientation of 55 µm particles due to paint application stress, resulting in higher $S_d$.

Notes:

## Poster 18: Small amplitude oscillatory rheometry for optimization of coatings formulations

### Dr. Aleksei Kostitsyn, YarLI Paint Research Institute

The rheological properties of coatings play an important role both in their production and in their application to the substrate. Such properties as delamination resistance, suitability for application with specific methods, flow, sag resistance, etc., depend on coating viscosity. Moreover, rheological properties of coatings are also indicative of many parameters necessary for coating formulation.
State-of-the-art rheometry can measure dynamic viscosity of coatings at different shear conditions from the state of rest to high shear rates. The method of Small Amplitude Oscillatory Shear Rheometry (SAOS) is used to study coating viscosity in the state of rest. In this method the rotor oscillation amplitude is very small and it doesn't destroy the sample structure. As a result we can measure the storage (G') and loss moduli (G"), the complex viscosity ($|\eta|$) and the loss factor ($\tan\delta$) .[1]
Measuring these values in the state of rest helps to monitor the changes in the coating structure resulting from specific actions (dispersant addition, curing, structuring due to rheological agents, etc.).
It is demonstrated that SAOS can be effectively used for selection of type and amount of dispersant for pigment pastes using the complex viscosity minimum on the concentration curve. SAOS can measure coating viscosity at low shear rates which is difficult to do in rotational mode due to transient viscosity peak [1].
It was shown that when the 2K PUR coatings are mixed with the hardener and then stored in a closed container, initially the viscosity goes up due to the loss modulus, but at a certain moment the storage modulus starts to increase abruptly as a result of formation of cross-linked network. This fact helps to determine the pot life of the 2K products more accurately.
We also present a method to determine the reaction onset temperature for polyester melamine systems using SAOS. The temperature of catalyst deblocking and reaction onset is determined by the abrupt increase of storage modulus and complex viscosity with temperature increase.

Notes:

## Poster 19: The effect of pigment surface modification on film formation of coatings

### Dr. Nataliia Skopintseva, YarLI Paint Research Institute

The influence of pigments and fillers on the formation of polymeric matrices, in particular those based on epoxy oligomers, has been yet insufficiently studied, especially due to significant broadening of the pigment brand range and of the existing methods of surface modification, especially for titanium dioxide.

To characterize the acid-base balance of titanium dioxide surface we measured the concentration and the strength of acid and base sites on it. To study the cross-linking agent adsorption on titanium dioxide surface one can calculate the minimum possible landing site, independent on the nature of adsorbing surface, which is limited not by the number of active sites on the adsorbing surface but rather by the chemical structure of adsorbate molecules.

Chemical and physical transformations during the polymer body formation start long before the coating is applied to the surface. First of all, one should note the influence of degree of diamine branching on the cross-linking rate. Although stechiometricity is not greatly impaired through hardener adsorption, this still affects the duration of the period from the onset of reaction between oligomer and cross-linking agent to the stage of three-dimensional network formation, as well as the final conversion level during curing of pigmented compositions.

During coating film formation the critical factor is not only the cross-linking rate, but rather the density of polymeric network, providing barrier effect, that is very important for protective properties of the coating. Another significant influence on the protective properties is the filler particle fineness that reduces the coating porosity and the internal stresses arising during coating formation.

Stability of dispersions containing titanium dioxide and an organic pigment depends on the particle charge signs that are a matter of the type of pigment surface modification.
The result of this study was the development of recommendations for the selection of titanium dioxide for different types of cross-linking agents, as well as for the period of pigmented composition storage time before application.

Notes:

## Poster 20: Synthesis and characterization of covalently bonded stabilizer

### Ms. Zuzana Nádvorníková, University of Pardubice

This paper deals with the stabilization of polymers by covalently bound antidegradants. Most manufacturers stabilize their products by adding stabilizers which may be eluted from the matrix during application, and the product rapidly degrades. [1] The concentration of low molecular weight commercially antidegradants added in a polymer can decrease with time, mainly due to extreme conditions such as high temperatures, exposure to liquids that cause leaching or evaporation of small molecules or mechanical stress. [2] This is not only environmentally unfriendly, but also uneconomical. In this case we focused on stabilizers covalently bonded due to a functional group.

The prepared systems are characterized by selected analytical and thermal methods, $^{1}H$ and $^{13}C$ NMR, gel permeation chromatography. Furthermore, the systems were subjected to accelerated aging in QUV panel, and then were evaluated from the point of view of hardness, gloss, adhesion, thickness and measurements using a spectrophotometer.

Notes:

## Poster 21: High Performance Dispersants for Economy Paints

### Dr. Anton Van Dyk, The Dow Chemical Company

*Co-Authors: Lidaris San Miguel Rivera Caroline Hinson, Sandy Umphrey, Trevor Ewers, Tom Kalantar, Mark Langille*

Economy or high PVC (pigment volume concentration) paints constitute about half of the volume of waterborne architectural coatings used globally, and are particularly important in Asian Pacific (AP), Latin American (LA), and European, Middle Eastern and African (EMEA) markets. These paints are predominantly composed of more economical extenders to reduce costs using small amounts of binder and $TiO_2$. As a result, high PVC paints generally suffer from poor wash resistance, durability and wet hiding. There is a need in this market for lowering the formulation cost with equal or improved film properties such as wash resistance, opacity and color acceptance among others. Dispersants play a key role in high PVC formulations to help disperse and stabilize pigments and extenders along with influencing many coatings properties. Typically, low cost polyacid dispersants are used in this market segment. However, premium dispersants have been found to improve many coating properties including hiding, wash resistance, and scrub resistance. Premium hydrophobic copolymer dispersants provide the formulator with additional latitude to vary the extender package (particle sizes and extender types) and the PVC to provide cost effective more sustainable solutions to performance targets. This work focuses on the development of a new dispersant with improved washability in high PVC formulations. An 80 PVC styrene-acrylic paint with HEC thickener was used as a screening formulation. The washability performance was assessed using a Gardener Scrub machine. The coatings were drawdown on Leneta scrub charts using a two inch wide six mil Bird Bar. New dispersant compositions with improved washability performance were identified. We have explored how dispersant composition affects key coating performance properties in this high PVC formulation space. An analysis of these results is presented and we discuss how dispersant interactions with the formulation can be selected to maximize cost effective opacity, wash and scrub resistance performance.

Notes:

## Poster 22: A Non-Contact, Multi-Layer Coating Thickness Sensor

### Dr. Robert May, TeraView

We present a non-contact, multi-layer coating thickness sensor based on the use of pulsed terahertz light, with operating principles similar to that of ultrasound thickness gauges but without the need for a coupling fluid. The coated surface is illuminated from a short distance with a focused, broadband pulse of terahertz light and the returned signal, containing reflections from the interfaces between adjacent coating layers, is analysed to determine the thickness of each layer. The initial application area is the automotive paint process to report thicknesses of individual paint layers after baking, and for which successful operation has been demonstrated on all basecoat types encountered: solid, pearlescent and metallic; as well as on powder coatings. For qualification of the newly developed technology, point measurements made using the terahertz thickness sensor are compared against those obtained from standard thickness sensing modalities: pulse-echo ultrasound, electromagnetic gauge and optical microscopy.

Notes:

## Poster 23: Optimization of a biobased 2K-PU coating made from a polyhydroxylated olive-oil

**Ms Sophie Yam, Laboratoire de Peintures et Vernis**

*Co-Authors: Simon Deroover, Dr. Catherine Dekerckheer*

Ecology and sustainable development are becoming a major concern in the field of coatings. Conventional 2K polyurethane coatings are known to have high efficiency but are mainly petroleum based.

The goal of this study is to increase the biobased content of a 2K-PU coating by using a polyhydroxylated olive oil, produced by an innovative synthesis that aims to reduce the ecological footprint in comparison with classical synthesis routes.

In order to obtain a coating which meets industrial performance criteria such as adhesion, hardness and chemical, mechanical and corrosion resistance, etc. concurrently with a reduced number of experiments, the use of design of experiments was chosen.

A Scheffe's mixture design has been used to establish mathematical models between the components of the coating and their specifications which allowed to optimize the coating formulation.

Comparative results of our optimized biobased formulation with a conventional one will be presented.

Notes:

## Poster 24: Preliminary studies on synthesis and characterization of silicone-acrylic/styrene dispersions

**Ms. Izabela Ofat-Kawalec, Industrial Chemistry Research Institute**

*Co-Authors: Janusz Kozakiewicz, Joanna Trzaskowska, Marcin Borys*

Aqueous silicone-acrylic/styrene dispersions with hybrid (core-shell) particle structure designed as binders for breathable architectural paints with enhanced hydrophobicity were synthesized via three different methods: (a) emulsion polymerization of silicone monomers in aqueous acrylic-styrene dispersion, (b) emulsion polymerization of acrylic/styrene monomers in aqueous silicone resin dispersion and (c) emulsion copolymerization of acrylic/styrene monomers and silicone monomers. Characterization of the resulting hybrid dispersions and of films and coatings produced from them revealed substantial differences in dispersions particle structure (TEM) as well as in film and coating properties, including Tg, hardness, water resistance, adhesion and surface free energy.

Notes:

## Poster 25: Improving corrosion protection through Mica


### Mr Mark Smit, LKAB Minerals

*Co-Author: Rob Lammertink*


Lamellar barrier fillers are known to improve the anti-corrosive performance of two-component epoxy type of coatings. The differences in the corrosion protective performance of Muscovite and Phlogopite Mica compared to Micaceous Iron Oxide, differing in particle size (distribution), aspect ratio and chemical composition have been evaluated. This has been done by comparing the recently introduced Electrochemical Impedance Spectroscopy (EIS) with the traditional and often certified method of Salt Spray Testing (SST). EIS evaluation proves to be a convenient tool for fast screening of the anti-corrosive behavior of coatings, confirmed by SST. Lamellar fillers, especially selected grades of Mica, increase the anti-corrosive performance of two component epoxy coatings over nodular fillers. In particular, Micronized Mica with a well-defined moderate particle size and narrow particle size distribution is extremely well suited for anti-corrosive properties.

Notes:

## Poster 26: The effect of new bio-based resins from plant oils on properties of architectural coatings


### Ms. Anna Ślusarczyk, Institute for Engineering of Polymer Materials and Dyes

*Co-Authors: A. Slusarczyk, M. Zubielewicz, A. Krolikowska*


Plant oils offer many advantages for coatings industry: renewability, world-wide availability, relatively low prices, low toxicity, inherent biodegrability and high purity. They can be transformed to a large variety of monomers and polymers for many different applications. In our work we have studied 1-component aliphatic polyurethane dispersions (PUDs) synthesized according to a two-step approach: first polymerization of vegetable oil-based polyols with diisocyanate and DMPTA followed by phase inversion in water. PUDs contain a hard segment and soft segment in polymer backbone with polymeric diols as soft segments and urethane and urea groups as hard segments [1,2]. The aim of our work was to investigate PUDs with different molecular structures (differences in hard and soft segments content), building blocks from different vegetable oils and chain extenders: two aliphatic diisocyanates with one or two cyclohexane groups. The PUDs were also examined by DMA.

The performance of bio-based PU dispersions as potential binders for coatings was evaluated. Film forming properties, compatibility with alkyd/aliphatic polyurethane copolymer and urethanised alkyd resin and compatibility with typical waterborne paints' additives were studied. Properties significant for architectural coatings were evaluated: hardness, wet scrub resistance, water vapor and $CO_2$ permeability and weathering resistance. Effect of nanoparticles, such as silica and silver, on mechanical properties and antimicrobial resistance were investigated.

The suitability of the tested biopolymers as binders depends strongly on the biopolymer composition and proportions of its components – soft and hard segments, as well as hydrophilic segment incorporated via DMPTA.

Biopolymer containing the smallest amount of H12MDI and DMPTA showed negative effect on the stability of the formulation and on the film formation. Differences in biopolymer behavior related to the polyol type used were also observed.

Catalytic effect of some bio-based polymers on alkyd/aliphatic polyurethane copolymer and urethanised alkyd resins was observed and thus eliminates the need of a drying agent.

The addition of nano-$SiO_2$ improves the mechanical properties of coatings and their weathering resistance, the influence of nano-Ag was comparable to that of organic biocides.


**Acknowledgments**

The research was carried out under the project titled CORNET/4/15/2013 Cost-effective biopolymer interior coatings (BioCoat) funded by the NCBiR.


Notes:

## Poster 27: Synthesis of Biobased-Poly(lactic acid) Acrylate resin for UV-Curable Coating Applications

**Mr. Nantana Jiratumnukul, Chulalongkorn University**
*Co-Author: Apinya Musidanga*

The objective of this research is to synthesize poly(lactic acid) (PLA) based acrylate to use as an oligomer in UV-curable coating applications. UV-curable process is widely used in paint, ink and adhesive industries due to its rapid curing speed, low energy consumption, high efficiency and low volatile organic compounds (VOCs). PLA was glycolyzed by ethylene glycol (EG) at 190oC for 90 minutes. The obtained glycolyzed PLA was then reacted with methacrylic anhydride (MAAH) to provide PLA acrylate oligomer. The PLA acrylate oligomer was used in coating formulations with various types and amounts of reactive diluents. Physical properties of cured coating film were investigated such as pencil hardness, gloss, chemical resistance and haze. The results showed that the UV-curable coating can be prepared from bio-base materials for environmental friendly concern. In addition, the coating film showed good physical properties.

Notes:

**Additional Notes**

# AUTHOR INDEX

**European Technical Coatings Congress ( ETCC )**
c/o Vicentz Network GmbH & Co. KG
Plathnerstr. 4c
30175 Hannover
Germany

ISBN 978-1-5108-2548-2

9 781510 825482

CURRAN ASSOCIATES INC.
proceedings.com